Molecular Biology Intelligence Unit

Protein Transport into the Endoplasmic Reticulum

Richard Zimmermann, PhD
Competence Center for Molecular Medicine
Saarland University
Homburg, Germany

Landes Bioscience
Austin, Texas
USA

Protein Transport into the Endoplasmic Reticulum

Molecular Biology Intelligence Unit

Landes Bioscience

Printed in the USA.

Please address all inquiries to the publisher:
Landes Bioscience, 1002 West Avenue, Austin, Texas 78701, USA
Phone: 512/ 637 6050; Fax: 512/ 637 6079
www.landesbioscience.com

The chapters in this book are available in the Madame Curie Bioscience Database.
http://www.landesbioscience.com/curie

ISBN: 978-1-58706-326-8

Library of Congress Cataloging-in-Publication Data

Protein transport into the endoplasmic reticulum / [edited by] Richard Zimmermann.
p. ; cm. -- (Molecular biology intelligence unit)
Includes bibliographical references and index.
ISBN 978-1-58706-326-8
1. Endoplasmic reticulum. 2. Proteins--Physiological transport. I. Zimmermann, Richard, Ph. D. II. Series: Molecular biology intelligence unit (Unnumbered : 2003)
[DNLM: 1. Endoplasmic Reticulum--metabolism. 2. Protein Transport--physiology. 3. Proteins--metabolism. QU 350 P967 2009]
QH603.E6P76 2009
571.6'5--dc22

2009022305

Dedication

To Maria and Julia

About the Editor...

RICHARD ZIMMERMANN is Professor of Biochemistry and Molecular Biology at the Medical School of Saarland University, Homburg, Germany. His research interests fall within the fields of intracellular protein transport and molecular chaperones, with an emphasis on the role of chaperones and co-chaperones in protein transport into the mammalian endoplasmic reticulum. He received his PhD from the "Georg-August"-University in Göttingen, Germany, and was Postdoc at the Molecular Biology Institute of the University of California at Los Angeles, USA. Before his current appointment he was Assistant Professor at the "Ludwig-Maximilians"-University in Munich and Associate Professor at the "Georg-August"-University of Göttingen, both Germany. Since 2005 he heads the Competence Center for Molecular Medicine at Saarland University (KoMM).

CONTENTS

EDITOR

Richard Zimmermann
Competence Center for Molecular Medicine
Saarland University
Homburg, Germany
Email: bcrzim@uks.eu
Chapter 9

CONTRIBUTORS

Deborah E. Aronson
Department of Anatomy
and Structural Biology
Albert Einstein College of Medicine
Bronx, New York, USA
Email: drussel@aecom.yu.edu
Chapter 3

Julia Benedix
Medical Biochemistry
and Molecular Biology
Saarland University
Homburg, Germany
Email: bcjben@uks.eu
Chapter 6

Nica Borgese
CNR Institute for Neuroscience
and Department of Pharmacology
Milano, Italy
and
Department of Pharmacobiological
Science
University of Catanzaro
Catanzaro, Italy
Email: n.borgese@in.cnr.it
Chapter 8

Silvia Brambillasca
CNR Institute for Neuroscience
and Department of Pharmacology
University of Milano
Milan, Italy
and
Department of Experimental Diabetology
School of Medicine
Dresden University of Technology
Dresden, Germany
Email: silviabrambi@hotmail.com
Chapter 8

Sara Colombo
CNR Institute for Neuroscience
Milano, Italy
Email: s.colombo@in.cnr.it
Chapter 8

Benedict C.S. Cross
Department of Life Sciences
University of Manchester
Manchester, UK
Email: benedict.cross@manchester.ac.uk
Chapter 7

Bernhard Dobberstein
Zentrum für Molekulare Biologie
der Universität Heidelberg (ZMBH)
Heidelberg, Germany
Email: b.dobberstein@zmbh.uni-heidelberg.de
Chapter 1

Johanna Dudek
Medical Biochemistry
and Molecular Biology
Saarland University
Homburg, Germany
Email: bcjdud@uks.de
Chapter 6

Frank Erdmann
Department of Biology
and Chemistry
University Osnabrück
Osnabrück, Germany
Email: frank.erdmann@biologie.uni-osnabrueck.de
Chapter 4

Anke Harsman
Department of Biology
and Chemistry
University Osnabrück
Osnabrueck, Germany
Email: anke.harsmann@biologie.uni-osnabrueck.de
Chapter 4

Stephen High
Department of Life Sciences
University of Manchester
Manchester, UK
Email: stephen.high@manchester.ac.uk
Chapter 7

Alf Honigmann
Department of Biology
and Chemistry
University Osnabrück
Osnabrück, Germany
Email: alf.honigmann@biologie.uni-osnabrueck.de
Chapter 4

Carolin Jalal
Medical Biochemistry
and Molecular Biology
Saarland University
Homburg, Germany
Email: bccjal@uks.eu
Chapter 6

Katja Kapp
Zentrum für Molekulare Biologie der
Universität Heidelberg (ZMBH)
Heidelberg, Germany
Email: k.kapp@zmbh.uni-heidelberg.de
Chapter 1

Marius K. Lemberg
Zentrum für Molekulare Biologie der
Universität Heidelberg (ZMBH)
Heidelberg, Germany
Email: m.lemberg@zmbh.uni-heidelberg.de
Chapter 1

Linda Müller
Medical Biochemistry and Molecular
Biology
Saarland University
Homburg, Germany
Email: bclmue@uks.eu
Chapter 9

Christopher V. Nicchitta
Department of Cell Biology
Duke University Medical Center
Durham, North Carolina, USA
Email: c.nicchitta@cellbio.duke.edu
Chapter 2

Marco Righi
CNR Institute for Neuroscience
and Department of Pharmacology
University of Milano
Milan, Italy
Email: m.righi@in.cnr.it
Chapter 8

Sabrina Schrempf
Zentrum für Molekulare Biologie der
Universität Heidelberg (ZMBH)
Heidelberg, Germany
Email: s.schrempf@zmbh.uni-heidelberg.de
Chapter 1

Erik L. Snapp
Department of Anatomy
and Structural Biology
Albert Einstein College of Medicine
Bronx, New York, USA
Email: esnapp@aecom.yu.edu
Chapter 3

Samuel B. Stephens
Sarah W. Stedman Nutrition
and Metabolism Center
Department of Pharmacology
and Cancer Biology
Duke University Medical Center
Duke University
Durham, North Carolina, USA
Email: sbs8@duke.edu
Chapter 2

Colin J. Stirling
Department of Life Sciences
University of Manchester
Manchester, UK
Email: colin.stirling@manchester.ac.uk
Chapter 5

Richard Wagner
Department of Biology
and Chemistry
University Osnabrück
Osnabrück, Germany
Email: wagner@uos.de
Chapter 4

Barrie M. Wilkinson
Department of Life Sciences
University of Manchester
Manchester, UK
Email: barrie.m.wilkinson@manchester.ac.uk
Chapter 5

Bernd Wullich
Urologische Klinik
Universitätsklinikum Erlangen
Erlangen, Germany
Email: bernd.wullich@uk-erlangen.de
Chapter 9

PREFACE

Protein transport into the endoplasmic reticulum (ER) is just one aspect of the general cell biology topic of intracellular protein sorting. This larger picture also includes protein transport into other organelles of the eukaryotic cell (chloroplasts, mitochondria, nucleus, peroxisomes), protein export from bacteria, vesicular transport that delivers to its final destination most of what has been transported into the ER, and protein export from the ER that is associated with protein degradation (termed ERAD). Over the years, protein transport into the ER also has become part of the quest to understand the various roles of molecular chaperones under non-stress conditions and of the ribosomal tunnel exit as the decisive site for molecular triage of nascent polypeptide chains (Table 1).

Protein transport into the ER is a first step in the biogenesis of most extracellular and many organellar proteins of eukaryotic cells (ER, ER-Golgi intermediate compartment, Golgi, endosome, lysosome). In addition, integration into the ER membrane is the initial step in the biogenesis of many membrane proteins (plasma membrane, nuclear envelope, peroxisomal membrane). Typically, protein transport into the ER involves amino terminal signal peptides in the precursor proteins and a transport machinery (Table 2). Two mechanisms can be distinguished that differ in their relationship to translation (termed co- and posttranslational mechanisms) and with respect to the relevant cytosolic components (ribosome plus signal recognition particle [SRP] versus cytosolic heat shock protein 70 [Hsp70] plus Hsp40). The two mechanisms merge at the ER membrane, specifically at the Sec61 complex that comprises α-, β-, and γ-subunits. Transport can be divided into three stages: targeting, membrane insertion, and completion of translocation. Targeting or membrane association of precursors in cotranslational transport involves the ribosome and SRP, and their receptors on the ER surface (the SRP and ribosome receptors). In yeast, targeting in posttranslational transport of pre-secretory proteins involves a heterotrimeric complex of membrane proteins, comprising Sec62p, Sec71p (Sec66p), and Sec72p (Sec67p) (Table 1). In yeast and in mammals, membrane insertion occurs at the level of the Sec61 complex and involves additional components, such as the ER-lumenal chaperone BiP or Kar2p and its co-chaperone (Sec63) and nucleotide exchange factors. The same is true for completion of translocation. The terms translocon and protein translocase are alternatively used for the Sec61 complex plus its partner proteins, depending on whether the activity is visualized as passive or active, respectively. Between membrane insertion and completion of translocation, the precursors are typically processed by the signal peptidase complex (SPC) and/or oligosaccharyl transferase (OST). Processing by SPC or OST, however, is not a prerequisite for transport into the ER. Posttranslational transport of tail-anchored membrane proteins into the ER involves an alternative cytosolic complex (the transmembrane domain-recognition complex [TRC/GET]) and its receptor on the ER surface (TRC/GET receptor) for membrane targeting.

It is noteworthy in this context that protein export from the bacterial cytosol is mechanistically related to protein transport into the ER. In fact, there are many evolutionarily related transport components present in the bacterial plasma membrane and the ER membrane (Table 1), and the signal peptides are very similar in both types of organisms and in some cases even interchangeable. However, there is at least one fundamental difference: In bacteria, transport is energized on the cis or cytosolic face of the membrane (by SecA), and transport into the ER is energized on the trans or ER lumenal side of the membrane (by Kar2p and BiP).

Typically, protein transport into the ER is followed by folding and assembly of the transport substrates. Folding and assembly of proteins also may involve some of the above-mentioned molecular chaperones. After folding and assembly, the native proteins are delivered to their functional location by vesicular transport. In the case of misfolding or misassembly, the polypeptides are exported to the cytosol and delivered to the proteasome for degradation (ERAD). Protein export from the ER lumen to the cytosol may also be mediated by the Sec61 complex. Furthermore, protein export also involves some of the above-mentioned molecular chaperones.

Richard Zimmermann, PhD

Table 1. ***Ribosomal ligands (proteins and complexes) that are involved in folding, membrane transport, or membrane integration of polypeptides. TF, trigger factor; RAC, ribosome associated complex, NAC, nascent chain associated complex; SRP, signal recognition particle, nd, not detected. Note that plastidic Alb3 is an ortholog of mitochondrial Oxa1 and that these two proteins are related to eubacterial YidC.***

Compartment	Mammalia	Yeasts	Plants	Eubacteria	Archaea
Cytosol				**TF**	
	RAC	**RAC**	**RAC**		
		Ssb	nd		
	Hsp70L1	Ssz	nd		
	MPP11 (ZRF1)	Zuotin	ZUO1		
	NAC	**NAC**	**NAC**		**NAC**
	α	α (Egd2p)	α		α
	β	β (Egd1p)	nd		α
	SRP	**SRP**	**SRP**	**SRP**	**SRP**
	7SL RNA	scR1 RNA	7S RNA	4.5S RNA	7S RNA
	SRP54	Srp54p	SRP54	Ffh	SRP54
	SRP19	Sec65p	SRP19		SRP19
	SRP68	Srp68p	SRP68		
	SRP72	Srp72p	SRP72		
	SRP9	Srp7p	SRP9		
	SRP14	Srp14p	SRP14		
		Srp21p			
ER- / plasma membrane	**Sec61 complex**	**Sec61 complex**	**Sec61 complex**	**SecYEG complex**	**SecY complex**
	Sec61α	Sec61p	Sec61α	SecY	SecYα
	Sec61β	Sbh1p	Sec61β	SecG	SecYβ
	Sec61γ	Sss1p	Sec61γ	SecE	SecYγ
	ERj1		nd		
Mitochondrial inner membrane	**Oxa1** **Letm1**	**Oxa1** **Mdm38**	**OXA1** **MDM38**		

Table continued on next page

Table 1. Continued

Compartment	Mammalia	Yeasts	Plants	Eubacteria	Archaea
Chloroplasts			**cpTF**		
			cpSRP		
			cpSRP54		
			cpSRP43		
			cpSecYE complex		
			cpSecY		
			cpSecE		
			Alb3		

Table 2. Proteins and complexes that are involved in protein transport into the endoplasmic reticulum

Function	*Homo sapiens*	*Saccharomyces cerevisiae*
Targeting machinery I	**SRP**	**SRP**
	7SL RNA	scR1 RNA
	SRP72	Srp72p
	SRP68	Srp68p
	SRP54	Srp54p
		Srp21p
	SRP19	Srp65p
	SRP14	Srp14p
	SRP9	Srp7p
	SRP-receptor (SR)	**SRP-receptor (SR)**
	SRα (docking protein)	SRα
	SRβ	SRβ
Targeting machinery II	**TRC**	**GET**
	TRC40 (Asna1)	GET3 (Arr4)
	TRC-receptor	**GET-receptor**
		GET1
		GET2

Table continued on next page

Table 2. Continued

Function	*Homo Sapiens*	*Saccharomyces cerevisiae*
Cytosolic chaperones	**Hsc70**	**Ssa1p**
	Hdj2	**Ydj1p**
	Hsp40	
Protein translocase (Translocon)	**Sec61-complex**	**Sec61-complex**
	Sec61α1	Sec61p
	Sec61β	Sbh1p
	Sec61γ	Sss1p
	Sec62/63-complex	**Sec62/63-complex**
	Sec62 (HTP1, TLOC1)	Sec62p
	Sec63 (ERj2)	Sec63p (Ptl1p, Npl1p)
		Sec66p (Sec71p)
		Sec67p (Sec72p)
	trans-acting chaperone	**trans-acting chaperone**
	BiP (Grp78, HspA5)	Kar2p
	NEFs	**NEFs**
	Grp170 (Orp150, HYOU1)	Lhs1p (Ssi1p, Cer1p)
	Sil1 (BAP)	Sil1p
	Alternative 61	**Alternative 61 complex**
	Sec61α2	Ssh1p
		Sbh2p
		Sss1p
	Alternative 62/63	
	ERj1 (Htj1, DNAJC1)	
Auxiliary components	**RAMP4 (SERP1)**	
	TRAM1	
	TRAM2	
	PAT-10	
	Sec61β-IAP	
	TRAP complex (SSR)	
	TRAPα	
	TRAPβ	
	TRAPγ	
	TRAP δ	

Table continued on next page

Table 2. Continued

Function	*Homo Sapiens*	*Saccharomyces cerevisiae*
Modifying enzymes	**Signal peptidase complex**	**Signal peptidase complex**
	SPC25	Spc3p
	SPC22/23	Spc2p
	SPC21	
	SPC18	Sec11p
	SPC12	Spc1p
	Oligosaccharyl transferase	**Oligosaccharyl transferase**
	Ribophorin I	OSTα (Ost1p)
	Ribophorin II	OSTδ (Swp1p)
	OST48 (Wbp1)	OSTß (Wbp1p)
	STT3-A	Stt3p
	STT3-B	
	N33 (OST3)	OSTγ (Ost3p)
	IAP (OST6)	Ost6p
		OSTε (Ost2p)
		Ost4p
		OSTζ (Ost5p)
Additional ER-chaperones	**ERj3 (hsDj9, HEDJ, DnaJB11)**	**Scj1p**
	ERdj4 (ERj4, DnaJB9)	
	ERdj5 (ERj5, DnaJC10)	
	p58IPK (DnaJC3)	
		Jem1p
		Erj5p
	Calreticulin (CaBP3, ERp60)	
	Calnexin (p90, p88)	**Cne1p**
	Grp94 (CaBP4, ERp99)	
	Hsp47	

Acknowledgements

I am thankful to my scientific teachers Walter Neupert and William Wickner who taught me both how to do science and how to enjoy science. Equally important for my career was the continuous funding by the Deutsche Forschungsgemeinschaft (DFG). I am very grateful for this support. Furthermore, I want to thank my colleagues who contributed chapters to this book. Many of them I have had fruitful and enjoyable collaborations with for a number of years. Last but not least I am thankful to my secretary Christiane Schäfer-Dennert for editorial help.

Chapter 1

Post-Targeting Functions of Signal Peptides

Katja Kapp, Sabrina Schrempf, Marius K. Lemberg* and Bernhard Dobberstein*

Abstract

Signal sequences are N-terminal extensions of newly synthesized secretory and membrane proteins. They are usually 16 to 30 amino acid residues in length and comprised of a hydrophilic, usually positively charged N-terminal region, a central hydrophobic domain and a C-terminal region with the cleavage site for signal peptidase. Besides these common characteristics, signal sequences do not share sequence similarity and some are more than 50 amino acid residues long.

In eukaryotes, signal sequences direct the insertion of proteins into the membrane of the endoplasmic reticulum and are usually cleaved off by signal peptidase. The resulting signal peptides are presumably rapidly degraded, but some still have functions on their own. Here, we describe examples of post-targeting functions of membrane-integral signal peptides, of signal peptides released from the membrane into either the cytosol or endoplasmic reticulum lumen and of signal peptide fragments generated by intramembrane cleavage. Thus, signal peptides must be considered as an additional resource in the context of the function of secretory and membrane proteins.

Introduction

Signal sequences are N-terminal extensions of nascent polypeptide chains that mediate protein targeting to the membrane of the endoplasmic reticulum (ER).[1,2] They are on average 16 to 30 amino acid residues in length comprising a characteristic tripartite structure: (1) a hydrophilic, usually positively charged n-region, (2) a central hydrophobic h-region of 5-15 residues and (3) a c-region with the cleavage site for signal peptidase (SPase) (Fig. 1A).[3] The consensus cleavage site consists of amino acids with short side chains at the −1 and no charged amino acid residues at the −3 position.[4] Besides these common features, signal sequences can be quite different in sequence and in length, displaying extended n-regions or two hydrophobic regions (Fig. 1A).

When a signal sequence emerges from the ribosome, it is recognized by the signal recognition particle (SRP).[5,6] SRP retards elongation of the nascent polypeptide chain until the ribosome-nascent chain complex has docked onto the SRP receptor at the membrane of the ER[7]. Upon docking, the nascent chain is inserted into the ER translocon and polypeptide chain elongation resumes.[8,9] The translocon is a heterotrimeric protein complex consisting of the Sec61α, Sec61β and Sec61γ chains[10] (see chapter by Wagner et al). Membrane insertion of the signal sequence into the translocation channel proceeds in a loop-like fashion with the n-region

*Corresponding Authors: Marius K. Lemberg and Bernhard Dobberstein—Zentrum für Molekulare Biologie der Universität Heidelberg (ZMBH), DKFZ-ZMBH Allianz, Im Neuenheimer Feld 282, D-69120 Heidelberg, Germany.
Emails: m.lemberg@zmbh.uni-heidelberg.de, b.dobberstein@zmbh.uni-heidelberg.de

Protein Transport into the Endoplasmic Reticulum, edited by Richard Zimmermann

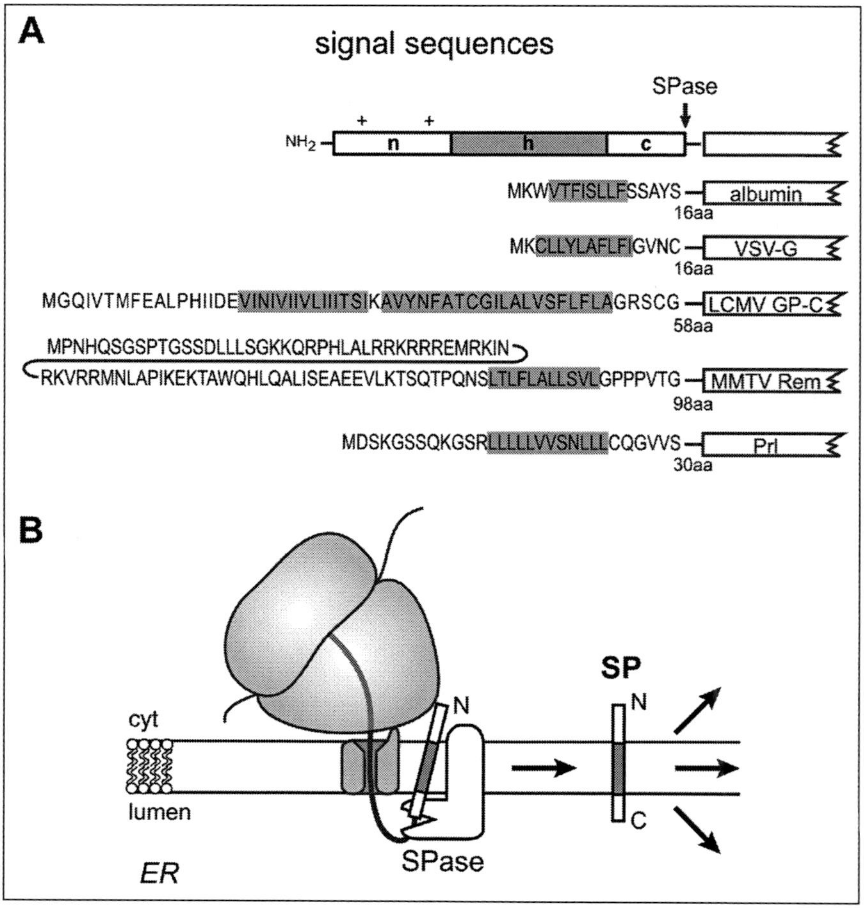

Figure 1. Signal sequence structures and signal peptide fates. A) Examples of differently sized signal sequences. Signal sequences can be as small as 16 amino acid residues but some are more then 50 amino acid (aa) residues in length. A characteristic feature of a signal sequence is its hydrophobic (h) region. Examples of minimal (albumin and VSV-G protein) and extended signal sequences (LCMV GP-C, MMTV Rem and prolactin (Prl) are shown. B) After insertion into the translocon in the ER membrane, signal sequences are usually cleaved off by signal peptidase (SPase) on the lumenal side of the ER membrane. The resulting signal peptides (SPs) initially accumulate in the ER membrane. Subsequently they can become degraded or can have functions as membrane-integrated peptides or as peptides released from the membrane either into the cytosol or the ER lumen.

remaining on the cytosolic side and the c-region emerging into the ER lumen (Fig. 1B).[1,11-13] Typically, signal sequences are cleaved off cotranslationally by SPase, a heterooligomeric complex consisting of five polypeptides (see ref. 14 for review).

Besides targeting proteins for insertion into the ER membrane, signal sequences affect interactions with the translocon or regulate membrane insertion efficiencies. These diverse targeting functions of signal sequences have been discussed previously.[15,16] Here we focus on functions of signal sequences after their insertion into the ER membrane and cleavage from their preprotein. Cleaved signal sequences, termed the signal peptides, are released from the translocation site into the lipid bilayer and spans the ER membrane in a carbonate-resistant manner like integral membrane proteins.[15,17-19] Subsequently signal peptides follow various routes ranging from persisting as a stable, membrane-integrated form to proteolytic clearance (Fig. 1B). The signal sequences of serum albumin and vesicular stomatitis virus G-protein (VSV-G) are among the shortest signal sequences (Fig. 1 A). They can be considered as minimal signal sequences that may only function in the targeting of the nascent preprotein for insertion into the ER membrane and may rapidly be degraded. Signal peptides for which post-targeting functions have been demonstrated, have extended n-regions or two h-regions (see Fig. 1A and below). These extended signal sequences have post-targeting functions, either as membrane-spanning peptide, as peptide released from the membrane or as signal peptide-fragment, generated by intramembrane proteolysis. Here we review currently known examples and discuss the emerging new paradigms of signal peptide function beyond protein targeting.

Signal Peptides That Have a Function as Membrane-Spanning Molecules

Transmembrane domains (TMDs) of integral membrane proteins typically consist of 20 to 25 residues, however even shorter peptide stretches such as the h-region of signal peptides can be sufficient to form stable TMDs.[20,21] To date all known examples of such stably membrane-integrated signal peptides are of viral origin and are derived from precursors of viral envelope glycoproteins.

Signal Peptides of Arenaviral Glycoproteins

Arenaviral glycoprotein C precursors (pGP-Cs) have unusual signal sequences.[22-24] They are longer than average signal sequences comprising 58 amino acid residues and contain an extended n-region including a myristoylation consensus site and two h-regions separated by a lysine residue (Fig. 1B and 2A). Signal sequences of arenaviral GP-Cs are highly conserved among all known family members reflecting an essential additional post-targeting function in the viral life cycle.[22,25-27] As a prototypic member we will discuss here the fate and functions of the signal peptide of the lymphocytic choriomeningitis virus (LCMV) glycoprotein[22,25,27] and point out differences to Lassa[23,28,29] and Junín viruses.[24,26,30]

The LCMV GP-C is synthesized as a precursor protein with an N-terminal signal sequence that is cleaved off by SPase during membrane insertion. The resulting signal peptide ($SP^{GP\text{-}C}$) remains membrane-anchored and associates with GP-C (Fig. 2B).[22-24,27] During intracellular transport, GP-C is proteolytically processed into GP1 and GP2. $SP^{GP\text{-}C}$ remains associated with the glycoproteins and is required for GP-C maturation into GP1 and GP2 and virus infectivity.[23,25-27,30] Besides having an extended n-region, $SP^{GP\text{-}C}$ has two h-regions (Figs. 1B and 2A). While one $SP^{GP\text{-}C}$ h-region is dispensable for targeting and membrane insertion, both h-regions are required for GP-C processing and transport to the cell surface.[25,27,29] Similarly, a substitution of the LCMV $SP^{GP\text{-}C}$ with an unrelated minimal (VSV-G) signal sequence prevents GP-C maturation. For the Lassa virus $SP^{GP\text{-}C}$, it has been shown that its function in glycoprotein maturation can also be provided in trans, that is, when synthesized separately from the glycoprotein.[28] The LCMV $SP^{GP\text{-}C}$ adopts an unusual topology in the ER membrane by exposing its n-region on the lumenal side of the membrane (Fig. 2B).[27] For the Junín virus $SP^{GP\text{-}C}$, however, an alternative topology with the n-region and c-region pointing towards the cytosol has been suggested.[31] In this way, the Junín virus $SP^{GP\text{-}C}$ is able to mask an ER-retrieval signal within the GP-C cytosolic domain, thus ensuring that only fully assembled GP complexes can leave the ER.[26] In contrast to Junín virus, the cytosolic portion of LCMV GP-C is not essential for the interaction with $SP^{GP\text{-}C}$.[27] Whether these reported differences indicate alternative topologies of $SP^{GP\text{-}C}$ during the viral life cycle or reflect evolutionary differences in glycoprotein maturation remains to be investigated. Myristoylation, the second unusual feature of the arenaviral $SP^{GP\text{-}C}$, is not needed for GP-C maturation but is proposed to be required for LCMV and Junín viral fusion with host cell membranes.[24,25,27,30] Thus, the individual regions of the arenaviral signal sequences play different roles in both, glycoprotein maturation and virus function.

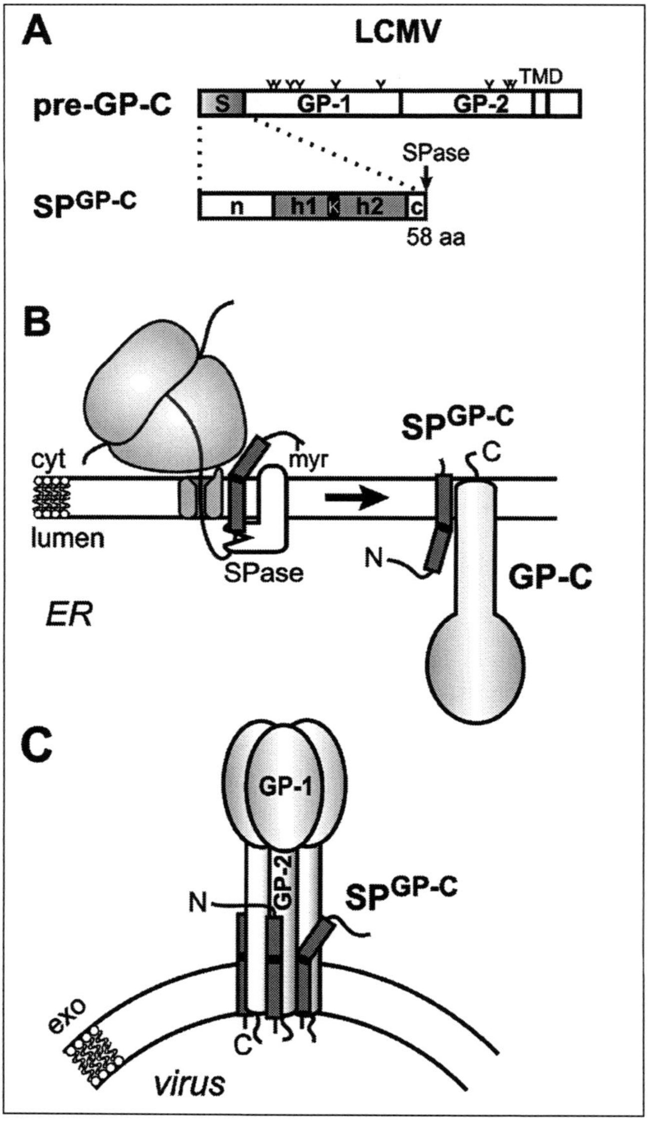

Figure 2. Signal peptides of arenaviral glycoproteins (GPs) are essential for glycoprotein maturation and viral infectivity. A) Outline of the GP-C precursor protein (pre GP-C) of the lymphocytic choriomeningitis virus (LCMV). Indicated are the signal sequence (S), the glycoprotein subunits GP-1 and the transmembrane domain (TMD) containing subunit GP-2 which are generated by site 1 protease cleavage in a late Golgi or post-Golgi compartment. Note that the signal peptide (SP^{REM}) has an unusual domain structure with an extended n-region and two h-regions. The charged lysine (K) is indicated. B) The signal sequence of the LCMV GP-C is myristoylated, inserted into the ER membrane and cotranslationally cleaved off by signal peptidase (SPase). The resulting signal peptide (SP^{GP-C}) associates with the GP-C. After trimerisation, the glycoprotein complex is transported to the plasma membrane and integrated into budding virus particles.[22,27] C) Schematic outline of the LCMV glycoprotein complex in the viral membrane. SP^{GP-C} as part of the viral glycoprotein complex is essential for viral infectivity.[35]

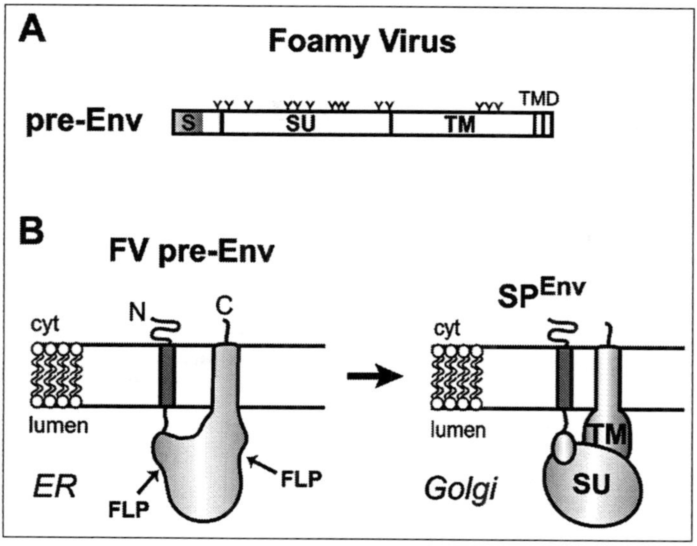

Figure 3. The signal sequence of Foamy virus (FV) Env protein is cleaved by a furin-like protease (FLP). A) Outline of the precursor of the Foamy virus envelope protein (pre-Env). Indicated are the signal sequence (S), the signal peptide (SP^{Env}), the glycoprotein subunit SU and the transmembrane domain (TMD) containing subunit TM.[32] B) The FV Env protein is synthesized as a precursor protein with an N-terminal signal sequence (formally representing a signal anchor), which is cleaved off in the Golgi by a furin-like protease (FLP).[33,34] The resulting signal peptide, SP^{Env} remains an essential component of the mature Env complex.

The Signal Peptide of the Foamy Virus Envelope Protein

The envelope proteins (Env or $gp130^{Env}$) of Foamy viruses are produced from a precursor protein (Fig. 3A).[32] Membrane insertion of the precursor is mediated by an extended signal sequence that is, however, not cleaved off by signal peptidase.[33,34] Thus, pre-$gp130^{Env}$ is anchored in the ER membrane by the hydrophobic domain of its signal sequence (formally defined as a signal anchor sequence) and the C-terminal TMD and exposes the N- and C-terminus to the cytosol (Fig. 3B). The cleavage of the N-terminal 126 amino acid residues of pre-$gp130^{Env}$ containing the signal anchor sequence occurs during intracellular transport by a furin-like protease.[33,34] A second cleavage by furin or a furin-like protease produces the SU and the membrane-integrated TM subunit. $SP^{FV\text{-}Env}$ is incorporated into viral particles where it interacts with the viral Gag protein.[35,36] Furthermore, $SP^{FV\text{-}Env}$ is found to be ubiquitinated and its ubiquitination is suggested to regulate the balance between viral and subviral particles released from infected cells.[37] Although not processed by SPase, the $SP^{FV\text{-}Env}$ demonstrates the potential to generate a functional membrane integral component from ER-targeting signal sequences.

Signal Peptides Released from the ER Membrane as Soluble Peptides

Although signal peptides initially accumulate in the ER membrane (Fig. 1B), most of them may not adopt a stable conformation in the lipid bilayer and may be released or extracted from the ER membrane.

The Release of the MMTV Derived Signal Peptide to the Cytosol

The mouse mammary tumour virus (MMTV) envelope protein (Env) is synthesized with an extended signal sequence of 98 amino acid residues.[38] Env-encoding transcripts contain an intron and are therefore only exported from the nucleus when special export factors are present (see below). When the intron is spliced out, the resulting mRNA encodes for the Rem protein (regulator of export/expression of MMTV mRNA).[39,40] Env and Rem share the same signal sequence, which has an extended n-region (1-80) containing a nuclear localization signal (NLS), (Fig. 4A).[41] When

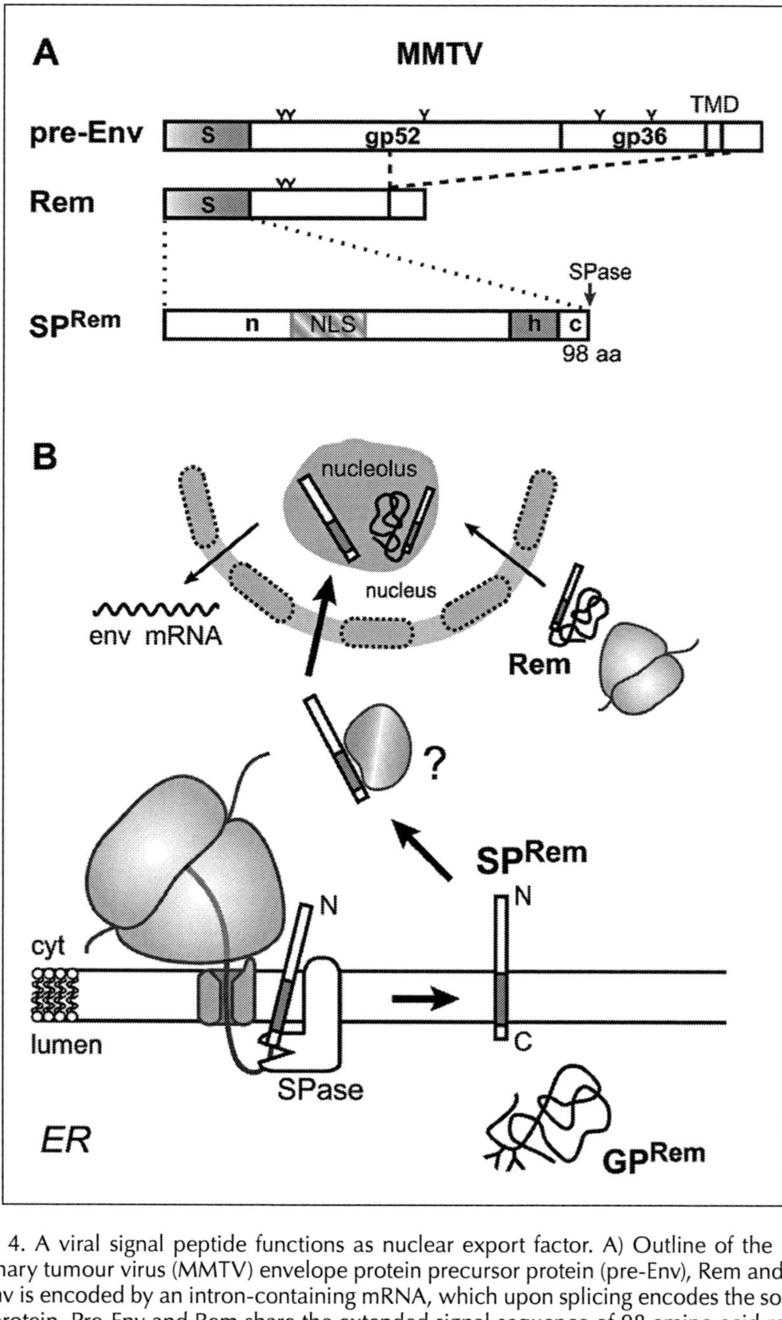

Figure 4. A viral signal peptide functions as nuclear export factor. A) Outline of the mouse mammary tumour virus (MMTV) envelope protein precursor protein (pre-Env), Rem and SP^{Rem}. Pre-Env is encoded by an intron-containing mRNA, which upon splicing encodes the so called Rem protein. Pre-Env and Rem share the extended signal sequence of 98 amino acid residues that contains a nuclear localisation sequence (NLS) in its N-terminal part.[39,40] B) Generation of SP^{Rem} and its transport to the nucleus. The signal sequence of Rem is cleaved cotranslationally by signal peptidase (SPase) leading to a secreted gylcoprotein (GP^{REM}). The resulting signal peptide^{Rem} is released from the membrane and transported to the nucleus where it functions in the nuclear export of the intron-containing mRNA, which encodes the Env protein.[42] In some cells full length Rem including the signal sequence can also directly be transported to the nucleus.

the Env precursors or Rem are inserted into the ER membrane, their signal sequences are cleaved off. The resulting signal peptides, $SP^{Rem/Env}$, initially accumulate in the ER membrane, then appear in the cytosol and, finally are found in the nucleus (Fig. 4B).[42] The membrane-translocated portions of Env and Rem become glycosylated and accumulate in the ER lumen as a Type I membrane protein and secretory protein respectively (Fig. 4B). Studies in a cell-free system have shown that SP^{Rem} is released from ER membranes in a time and temperature-dependent process and that this process is independent of a proteolytic processing step involving signal peptide peptidase (SPP) (see ref. 42 and below). In cells transiently expressing Rem, SP^{Rem} is found in the nucleus. In MMTV infected, murine mammary gland-derived cells and when overexpressed, Rem is not exclusively targeted to the ER but also directly by its uncleaved signal sequence transported into the nucleus. Thus, the presence of two signals, the ER-signal sequence and the NLS, within the N-terminus of Rem allows for a dual targeting of the precursor. For nuclear transport, the NLS was found to be essential. Nuclear $SP^{Rem/Env}$ as well as Rem function in the nuclear export of intron-containing viral transcripts.[39] Besides the NLS, $SP^{Rem/Env}$ contains a putative leucine-rich nuclear-export signal in its hydrophobic region.[42-44] The function of $SP^{Rem/Env}$ as nuclear export factor appears to be similar to the well-characterized function of human immunodeficiency virus (HIV) Rev protein.[39,44] The ability to generate such a regulatory factor from the ER-targeting signal sequence adds an intriguing layer to the complexity of the retroviral life cycle.

The Release of a Signal Peptide to the ER Lumen

The envelope glycoproteins of the alphaviruses, Semliki Forest Virus (SFV) and Sinbis Virus, are produced as part of a polyprotein that is sequentially processed cotranslationally by both viral and cellular proteases to yield the cytosolic nucleocapsid protein and the membrane proteins

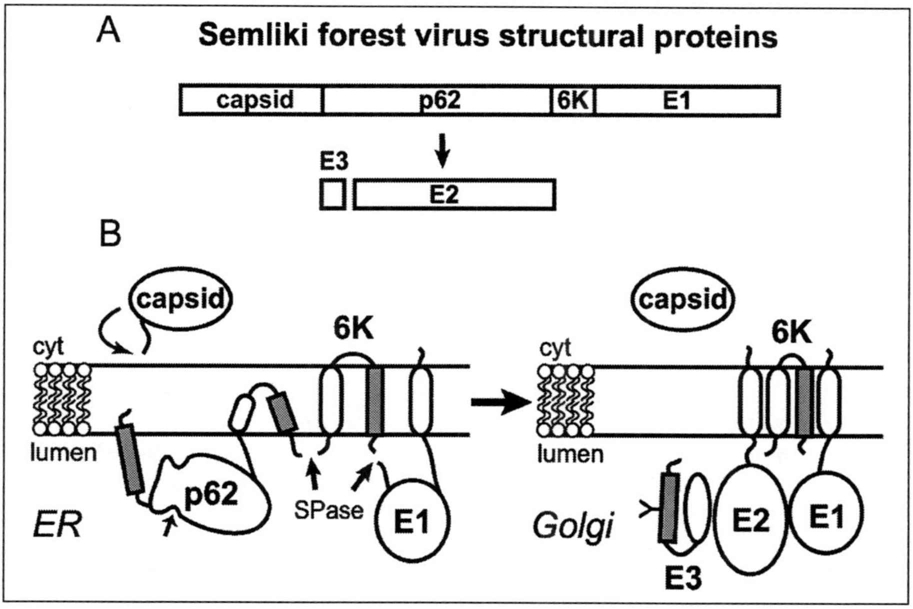

Figure 5. Release of a signal sequence to the ER lumen. A) Outline of the Semliki Forest virus (SFV) structural proteins. Indicated are the capsid protein, the glycoprotein subunit p62 (which gets further processed into E3 and E2), the 6K peptide and E1.[45,46] B) The SFV structural proteins are synthesised at the ER as a polyprotein. After autoproteolysis by the capsid protease, p62 is translocated into the ER lumen with its uncleaved signal sequence.[47] The signal sequence becomes glycosylated and two internal signal sequences then mediate membrane insertion of the 6K and E1 proteins and are cleaved off by signal peptidase (SPase). In the Golgi, the N-terminal 66 amino acid residues of p62 comprising the signal sequence are cleaved off. The resulting E3 protein remains part of the glycoprotein complex and promotes viral assembly and budding.[49]

p62, 6K and E1 (Fig. 5).[45,46] During cotranslational processing of the polyprotein precursor, the nucleocapsid protein is autoproteolytically cleaved and remains in the cytosol.[47] A stretch of apolar residues at the N-terminus of p62 functions as a signal sequence that targets p62 to the ER. After its membrane insertion the signal sequence of p62 is not cleaved off by SPase. Instead p62 is N-glycosylated within its signal sequence and subsequently cleaved in a post-Golgi compartment to generate E3 and E2 (Fig. 5).[45,48] The E3 peptide, essentially representing the signal peptide of p62, remains noncovalently associated with the E1-E2 spike glycoproteins and is found in mature virus particles. Replacement of E3 with an artificial signal sequence abolishes spike heterodimerisation and surface expression of E1.[49] E3 thus performs the function of a signal sequence and after cleavage the resulting peptide acts as soluble glycoprotein with an important role in the viral life cycle.[49]

In addition to the generation of this soluble, signal sequence-derived peptide, alpha viruses also produce a second membrane-integrated signal peptide, the 6K peptide. The coding region for this small 6 kDa membrane protein lies in between those of the p62 and E1 proteins (Fig. 5). The 6K protein results from two SPase cleavages and essentially consists of an N-terminal TMD and the cleaved signal sequence responsible for the insertion of the E1 protein.[46] Deletion of the 6K protein has revealed that it is required for efficient assembly of virus and is also involved in the viral budding process.[50-54]

Signal Peptide Processing and the Generation of Bioactive Signal Peptide Fragments

After cleavage by SPase some of the resulting signal peptides are further processed within their hydrophobic regions by SPP (signal peptide peptidase) (Fig. 6 and Table 1).[15,55] SPP is a member of the aspartic intramembrane proteases and is related to presenilin, the proteolytically active subunit of the γ-secretase complex.[56-58] SPP is localised to the ER while several SPP-like proteins are localized to different compartments of the cell.[59-62] Substrates thus far identified for SPP are all signal peptides or internal signal sequences of viral proteins that have been generated by SPase cleavage (Table 1).[55] Comparing signal peptides that are substrates with those that are not, revealed that helix-breaking residues within the h-region are required for efficient processing and flanking regions can affect processing. In order to be cleaved by SPP, substrate peptides or proteins have to be liberated from the nascent chain by SPase.[55] It was proposed that SPP substrates require flexibility in the lipid bilayer to expose an accessible peptide bond for intramembrane proteolysis, however, the exact substrate specificity is not completely understood.[63] Several signal peptides and internal signal sequences of viral polyproteins have been shown by now to be processed by SPP and for some of them specific functions of the generated fragments have either been proposed or demonstrated (Table 1).

The Signal Peptide Fragment of Preprolactin Targets Calmodulin

The signal peptide derived from SPase cleavage of the prolactin precursor (pre-Prl) is the first for which processing was detected. An in vitro test system was used in which ^{35}S-labelled nascent chains of pre-Prl were trapped in membrane vesicles derived from the rough ER.[17,18] After SPase cleavage of the signal sequence, the resulting SP^{Prl} is cleaved within its membrane-spanning h-region. An N-terminal fragment initially accumulates in the membrane but is liberated with time from the membrane into the cytosol.[17] In the cytosol, the SP^{Prl}-fragment was found to be associated with calmodulin in a Ca^{2+}-dependent manner suggesting a regulatory function.[18] It was proposed that the SP^{Prl} fragment may function in the feedback regulation of prolactin secretion. Similarly, the signal peptide of the HIV-1 Env glycoprotein was also found to be cleaved by SPP and to associate with calmodulin.[18] Binding of Env signal peptide fragment may contribute to the inactivation of calmodulin-dependent processes in HIV-infected T-cells.[64]

Signal Peptide Fragments of MHC Class I Molecules Signal to Natural Killer Cells

Classical MHC class I molecules present cellular peptides to T killer cells and thereby can report the presence of viral proteins in an infected cell. Natural killer (NK) cells recognize and

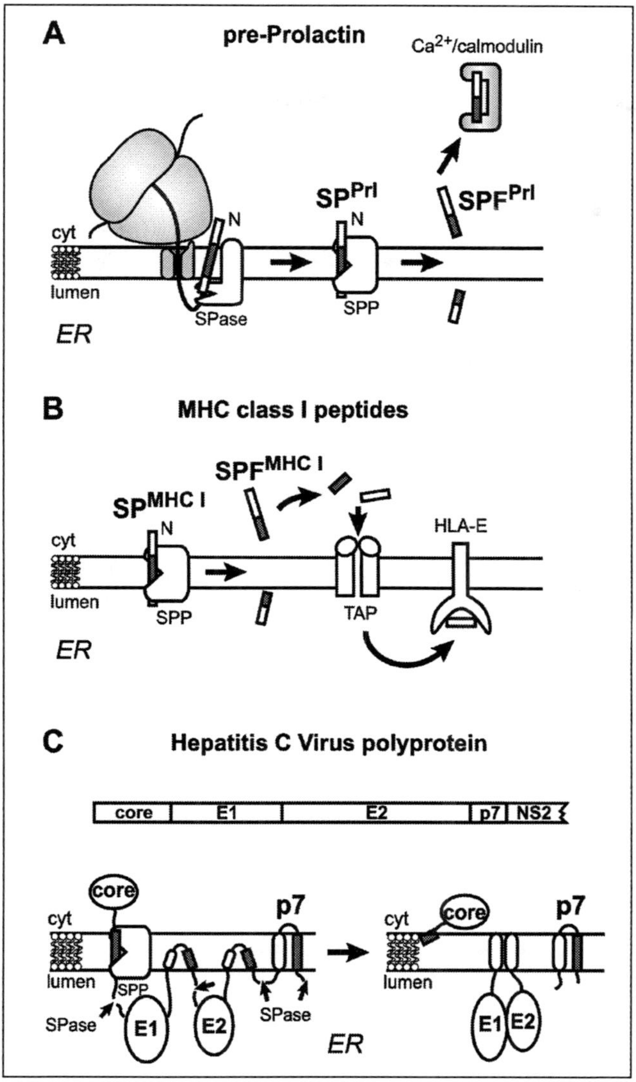

Figure 6. Generation of signal peptide fragments and their functions. A) The signal sequence of preprolactin (pre-Prl) is cleaved by signal peptidase (SPase) and the resulting SP^{Prl} is further processed by the intramembrane protease signal peptide peptidase (SPP). The N-terminal signal peptide fragment (SPF^{Prl}) is then released from the membrane and interacts with the signalling molecule calmodulin in a Ca^{2+}-dependent manner.[17,18] B) Signal peptides of classical MHC class I molecules ($SP^{MHC\ I}$) are processed by SPP and the resulting signal peptide fragments ($SPF^{MHC\ I}$) are trimmed by the cytosolic proteasome and upon TAP-transport by ER-lumenal proteases. A highly conserved nonamer binds to the nonclassical MHC class I molecule HLA E, which reports the appropriate biosynthesis of classical MHC class I molecules to natural killer cells.[19,66,67] C) The structural proteins of the hepatitis C virus (HCV) are synthesized as the N-terminal portion of the HCV polyprotein. The internal signal sequence between the core and E1 protein targets the nascent chain to the ER membrane. The HCV core, E1, E2 and p7 proteins are liberated from the polyprotein by consecutive cleavages by SPase.[71] Subsequently the immature core protein is processed by SPP in the membrane thereby enabling trafficking to cytosolic lipid droplets.[72-74]

Table 1. SPP substrates. Upon liberation by SPase, signal peptides of the following precursor proteins are processed by SPP. Known or suggested post-targeting functions and the bioactive fragments are listed.

Precursor Protein	Post-Targeting Function	Relevant Fragment	References
Preprolactin	Putative regulation of Ca^{2+}/ calmodulin-dependent processes	N-terminal signal peptide fragment	17,18
HIV pre-Env	Putative regulation of Ca^{2+}/ calmodulin-dependent processes	N-terminal signal peptide fragment	18
pre-HLA-A0301, pre-HLA-A2901	NK-cell dependent immune surveillance	Nonamer within the N-terminal signal peptide fragment binds to HLA-E	66,67,19
HCV polyprotein	N-terminal portion forms the viral nucleocapsid and can traffic to cytosolic lipid droplets. Cleavage by SPP is required for virus production	SPP cleaves within the internal signal sequence; different C-termini of the mature core protein had been suggested (between amino acid residues 171 and 179)	72,73,77
GB virus polyprotein	N-terminal portion forms the viral nucleocapsid and can traffic to cytosolic lipid droplets	SPP cleaves within the internal signal sequence	77
Pestivirus polyprotein	N-terminal portion forms the viral nucleocapsid	SPP cleaves within the internal signal sequence; the C-terminus of the mature core protein was assigned as alanine 255	89
Precalreticulin	No significant function known	Nonamer from the N-terminal signal peptide fragment has been observed on classical antigen presenting MHC class I molecules in a TAP-independent manner	55,90
pre-VSV-G	No significant function known		55

eliminate cells that lack MHC class I molecules thereby serving as an additional checkpoint in cellular immune surveillance. Among other mechanisms (reviewed by ref. 65), this recognition is mediated by a nonclassical MHC class I molecule, termed HLA-E, which presents nonameric peptide epitopes derived from the signal sequences of classical MHC class I molecules.[66] Signal sequences of the classical MHC class I molecules are highly conserved. They are cleaved off during insertion into the ER membrane by SPase and subsequently processed by SPP in the plane of the membrane. An N-terminal fragment of about 14 residues length is then released into the cytosol where it is trimmed by the proteasome (Fig. 6B).[19,67] After re-import into the ER lumen by the transporter associated with antigen processing (TAP), the peptide is loaded onto newly synthesized HLA-E molecules.[19] After transport to the cell surface, the HLA-E/peptide complex interacts with CD94/NKG2 receptors, resulting in the inhibition of NK cell activation.[68]

In an intriguing variation, human cytomegalovirus (HCMV) has evolved an alternative pathway leading to the generation of the HLA-E binding peptide. This is part of a complex immune evasion strategy down regulating MHC class I molecules (reviewed by ref. 68). To evade killing by NK cells, HCMV upregulates surface expression of HLA-E molecules. This is achieved by expression of a protein, gpUL40, which carries in its signal sequence an epitope that mimics an HLA-E ligand.[69,70] Generation of this epitope containing peptide does not require processing by SPP and HLA-E loading is independent of the TAP transporter.[68] This may point to a novel route by which SP^{UL40} fragments are generated and reach the HLA-E molecule in the ER lumen.

Intramembrane Processing of the HCV Core Protein Regulates Its Membrane Localisation

The structural components of Hepatitis C Virus (HCV) particles consist of the core protein, which forms the virus capsid, two envelope glycoproteins and a p7 protein (Fig. 6C).[71] These proteins are synthesized as the N-terminal portion of the HCV polyprotein at the host cell ER followed by the synthesis of the nonstructural proteins. An internal signal sequence between the core protein and E1 targets the nascent polypeptide to the ER for insertion of E1 into the membrane, where SPase cleaves the internal signal sequence, which then forms the C-terminal portion of the core protein. Further processing by SPP in the signal peptide h-region generates the mature form of the core protein.[72,73] This processing step is also required for subsequent trafficking of the core protein to cytosolic lipid droplets[72,74] and has recently been shown to be essential for viral propagation.[75-77] Not all signal sequences of the HCV polyprotein are substrate for SPP. Three further internal signal sequences mediate membrane insertion of the E2, p7 and the NS2 proteins. After cleavage by SPase, the resulting signal peptides become the TMDs of the E1, E2 and p7 proteins.[78] An interesting caveat here is that two of the signal peptides seem to flip across the membrane after the cleavage by SPase and expose their C-termini on the cytosolic side.[78,79] The two TMDs of the p7 protein are derived from a stop transfer sequence and a signal sequence. The p7 protein can form a heptameric ion channel that is blocked by the antiviral drug, Amantadine.[80,81] In summary, four signal peptides are functionally involved in the life cycle of HCV. One of these is processed by SPP and the other three function in the context of membrane-integrated proteins.

Conclusion and Perspectives

We have described here selected examples of signal peptides that perform distinct post-targeting functions in different cellular locations: in a membrane, in the cytosol/nucleus or in the ER lumen (see Fig. 7). Most examples are currently from viral proteins, thus it will now be interesting to see how common post-targeting functions of signal peptides derived from cellular proteins are. It is likely, that many signal peptides contain just the targeting information and are rapidly degraded. Little is known about signal peptide degradation, however, there are hints that SPP may play a role in this respect. The signal peptides of calreticulin, an ER lumenal lectin and the VSV-G protein are apparently substrates for SPP and the resulting fragments are expected to be nonfunctional.[55] Furthermore, SPP has been implicated in the degradation of misfolded TMD proteins[82,83] supporting the notion of SPP involvement in signal peptide degradation. Also

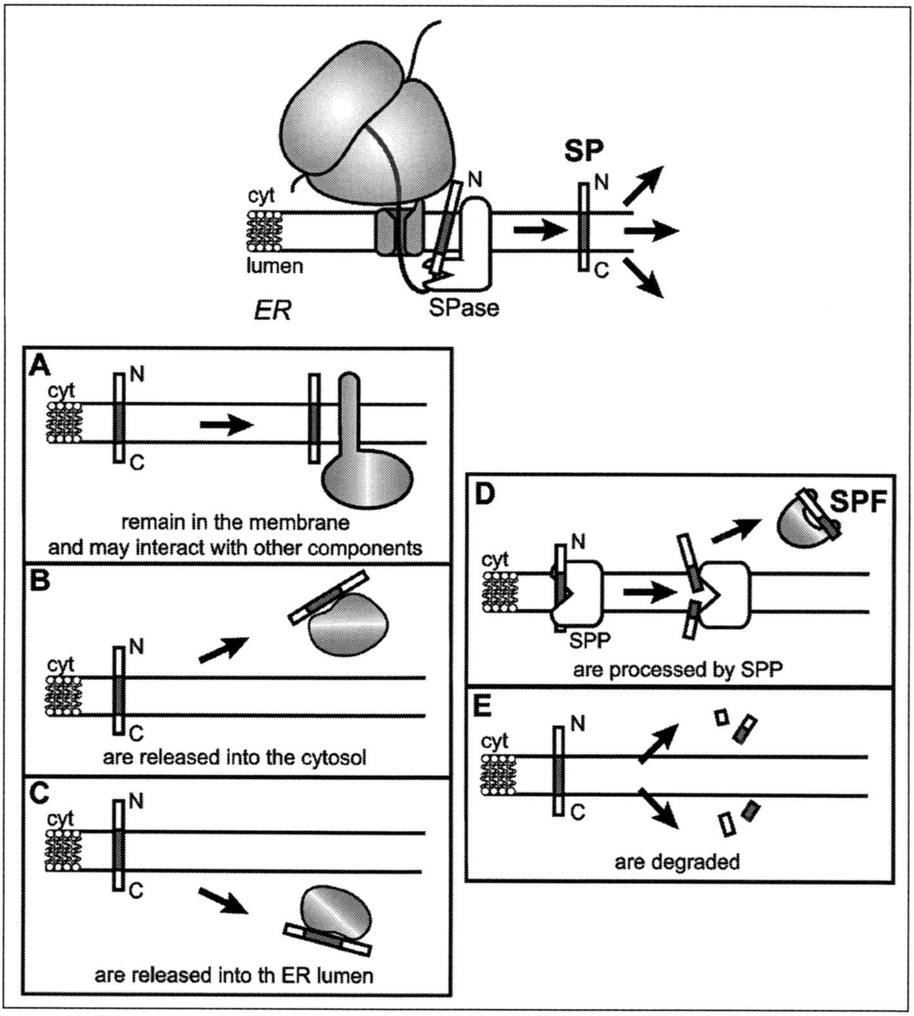

Figure 7. Summary of different fates and functions of signal peptides. A) Signal peptides can function as membrane integrated peptides (e.g., signal peptides of arenavirus and foamy virus glycoproteins). B) Signal peptides can be released from the membrane to the cytosol (signal peptide of MMTV pre-Env and Rem proteins). C) Signal peptides can be released from the membrane to the ER lumen (signal peptide of SFV). D) Signal peptides can be processed by SPP and the resulting fragments can have various functions (see Table 1). E) Directly after their cleavage from the precursor protein signal peptides may become degraded. This is probably true for most signal peptides but little is known about this process.

alternative degradation pathway may exist in which full-length signal peptides are ubiquitinylated and dislocated to the cytosolic 26S proteasome like misfolded membrane proteins.[84-87] Intriguingly, it has recently been shown that intracellular peptides similar to those generated by the proteasome have the potential to affect cell signalling,[88] implying an even bigger complexity of peptide mediated cellular processes. A detailed understanding of the molecular machinery involved in the turn over of signal peptides and the growing number of known post-targeting functions of signal sequences will likely give new insight into functions that signal peptides can perform after their cleavage from the parent protein.

Acknowledgements

We apologize to all those colleagues whose work has not been cited owing to space limitations. M.K.L. is supported by the Landesstiftung Baden-Württemberg and B.D. is supported by a grant from the Deutsche Forschungsgemeinschaft, SFB 638. We thank Stephen High and Martin Pool for critically reading of the manuscript and helpful comments.

References

1. Blobel G. Intracellular protein topogenesis. Proc Natl Acad Sci USA 1980; 77:1496-500.
2. Blobel G. Protein targeting (Nobel lecture). Chembiochem 2000; 1:86-102.
3. von Heijne G. Signal sequences. The limits of variation. J Mol Biol 1985; 184:99-105.
4. von Heijne G. A new method for predicting signal sequence cleavage sites. Nucleic Acids Res 1986; 14:4683-90.
5. Walter P, Gilmore R, Blobel G. Protein translocation across the endoplasmic reticulum. Cell 1984; 38:5-8.
6. Walter P, Johnson AE. Signal sequence recognition and protein targeting to the endoplasmic reticulum membrane. Annu Rev Cell Biol 1994; 10:87-119.
7. Walter P, Blobel G. Translocation of proteins across the endoplasmic reticulum III. Signal recognition protein (SRP) causes signal sequence-dependent and site- specific arrest of chain elongation that is released by microsomal membranes. J Cell Biol 1981; 91:557-61.
8. Meyer DI, Krause E, Dobberstein B. Secretory protein translocation across membranes-the role of the 'docking protein'. Nature 1982; 297:647-50.
9. Gilmore R, Walter P, Blobel G. Protein translocation across the endoplasmic reticulum. II. Isolation and characterization of the signal recognition particle receptor. J Cell Biol 1982; 95:470-7.
10. Rapoport TA. Protein translocation across the eukaryotic endoplasmic reticulum and bacterial plasma membranes. Nature 2007; 450:663-9.
11. Goder V, Spiess M. Topogenesis of membrane proteins: determinants and dynamics. FEBS Lett 2001; 504:87-93.
12. Evans EA, Gilmore R, Blobel G. Purification of microsomal signal peptidase as a complex. Proc Natl Acad Sci USA 1986; 83:581-5.
13. Shaw AE, Rottier PJM, Rose JK. Evidence for the loop model of signal sequence insertion into the endoplasmic reticulum. Proc Natl Acad Sci USA 1988; 85:7592-96.
14. Paetzel M, Karla A, Strynadka NC et al. Signal peptidases. Chem Rev 2002; 102:4549-80.
15. Martoglio B, Dobberstein B. Signal sequences: more than just greasy peptides. Trends Cell Biol 1998; 8:410-5.
16. Hegde RS, Bernstein HD. The surprising complexity of signal sequences. Trends Biochem Sci 2006; 31:563-71.
17. Lyko F, Martoglio B, Jungnickel B et al. Signal sequence processing in rough microsomes. J Biol Chem 1995; 270:19873-8.
18. Martoglio B, Graf R, Dobberstein B. Signal peptide fragments of preprolactin and HIV-1 p-gp160 interact with calmodulin. EMBO J 1997; 16:6636-45.
19. Lemberg MK, Bland FA, Weihofen A et al. Intramembrane proteolysis of signal peptides: an essential step in the generation of HLA-E epitopes. J Immunol 2001; 167:6441-6.
20. Sipos L, von Heijne G. Predicting the topology of eukaryotic membrane proteins. Eur J Biochem 1993; 213:1333-40.
21. von Heijne G. Membrane-protein topology. Nat Rev Mol Cell Biol 2006; 7:909-18.
22. Froeschke M, Basler M, Groettrup M et al. Long-lived signal peptide of lymphocytic choriomeningitis virus glycoprotein pGP-C. J Biol Chem 2003; 278:41914-20.
23. Eichler R, Lenz O, Strecker T et al. Signal peptide of Lassa virus glycoprotein GP-C exhibits an unusual length. FEBS Lett 2003; 538:203-6.
24. York J, Romanowski V, Lu M et al. The signal peptide of the Junin arenavirus envelope glycoprotein is myristoylated and forms an essential subunit of the mature G1-G2 complex. J Virol 2004; 78:10783-92.
25. Saunders AA, Ting JP, Meisner J et al. Mapping the landscape of the lymphocytic choriomeningitis virus stable signal peptide reveals novel functional domains. J Virol 2007; 81:5649-57.
26. Agnihothram SS, York J, Nunberg JH. Role of the stable signal peptide and cytoplasmic domain of G2 in regulating intracellular transport of the Junin virus envelope glycoprotein complex. J Virol 2006; 80:5189-98.

27. Schrempf S, Froeschke M, Giroglou T et al. Signal peptide requirements for lymphocytic choriomeningitis virus glycoprotein C maturation and virus infectivity. J Virol 2007; 81:12515-24.
28. Eichler R, Lenz O, Strecker T et al. Identification of Lassa virus glycoprotein signal peptide as a trans-acting maturation factor. EMBO Rep 2003; 4:1084-8.
29. Eichler R, Lenz O, Strecker T et al. Lassa virus glycoprotein signal peptide displays a novel topology with an extended endoplasmic reticulum luminal region. J Biol Chem 2004; 279:12293-9.
30. York J, Nunberg JH. Role of the stable signal peptide of Junin arenavirus envelope glycoprotein in pH-dependent membrane fusion. J Virol 2006; 80:7775-80.
31. Agnihothram SS, York J, Trahey M et al. Bitopic membrane topology of the stable signal peptide in the tripartite Junin virus GP-C envelope glycoprotein complex. J Virol 2007; 81:4331-7.
32. Lindemann D, Goepfert PA. The foamy virus envelope glycoproteins. Curr Top Microbiol Immunol 2003; 277:111-29.
33. Duda A, Stange A, Luftenegger D et al. Prototype foamy virus envelope glycoprotein leader peptide processing is mediated by a furin-like cellular protease, but cleavage is not essential for viral infectivity. J Virol 2004; 78:13865-70.
34. Geiselhart V, Bastone P, Kempf T et al. Furin-mediated cleavage of the feline foamy virus Env leader protein. J Virol 2004; 78:13573-81.
35. Lindemann D, Pietschmann T, Picard-Maureau M et al. A particle-associated glycoprotein signal peptide essential for virus maturation and infectivity. J Virol 2001; 75:5762-71.
36. Wilk T, Geiselhart V, Frech M et al. Specific interaction of a novel foamy virus Env leader protein with the N-terminal Gag domain. J Virol 2001; 75:7995-8007.
37. Stanke N, Stange A, Luftenegger D et al. Ubiquitination of the prototype foamy virus envelope glycoprotein leader peptide regulates subviral particle release. J Virol 2005; 79:15074-83.
38. Arthur LO, Copeland TD, Oroszlan S et al. Processing and amino acid sequence analysis of the mouse mammary tumor virus env gene product. J Virol 1982; 41:414-22.
39. Mertz JA, Simper MS, Lozano MM et al. Mouse mammary tumor virus encodes a self-regulatory RNA export protein and is a complex retrovirus. J Virol 2005; 79:14737-47.
40. Indik S, Gunzburg WH, Salmons B et al. A novel, mouse mammary tumor virus encoded protein with Rev-like properties. Virology 2005; 337:1-6.
41. Hoch-Marchaim H, Weiss AM, Bar-Sinai A et al. The leader peptide of MMTV Env precursor localizes to the nucleoli in MMTV-derived T-cell lymphomas and interacts with nucleolar protein B23. Virology 2003; 313:22-32.
42. Dultz E, Hildenbeutel M, Martoglio B et al. The signal peptide of the mouse mammary tumor virus Rem protein is released from the endoplasmic reticulum membrane and accumulates in nucleoli. J Biol Chem 2008; 283:9966-76.
43. Kutay U, Guttinger S. Leucine-rich nuclear-export signals: born to be weak. Trends Cell Biol 2005; 15:121-4.
44. Cullen BR. Nuclear mRNA export: insights from virology. Trends Biochem Sci 2003; 28:419-24.
45. Garoff H, Kondor-Koch C, Riedel H. Structure and assembly of alphaviruses. Curr Top Microbiol Immunol 1982; 99:1-50.
46. Liljestrom P, Garoff H. Internally located cleavable signal sequences direct the formation of Semliki forest virus membrane proteins from a polyprotein precursor. J Virol 1991; 65:147-54.
47. Melancon P, Garoff H. Processing of the Semliki forest virus structural polyprotein: role of the capsid protease. J Virol 1987; 61:1301-9.
48. Garoff H, Huylebroeck D, Robinson A et al. The signal sequence of the p62 protein of Semliki forest virus is involved in initiation but not in completing chain translocation. J Cell Biol 1990; 111:867-76.
49. Lobigs M, Zhao HX, Garoff H. Function of Semliki forest virus E3 peptide in virus assembly: replacement of E3 with an artificial signal peptide abolishes spike heterodimerization and surface expression of E1. J Virol 1990; 64:4346-55.
50. Liljestrom P, Lusa S, Huylebroeck D et al. In vitro mutagenesis of a full-length cDNA clone of Semliki forest virus: the small 6,000-molecular-weight membrane protein modulates virus release. J Virol 1991; 65:4107-13.
51. Loewy A, Smyth J, von Bonsdorff CH et al. The 6-kilodalton membrane protein of Semliki forest virus is involved in the budding process. J Virol 1995; 69:469-75.
52. McInerney GM, Smit JM, Liljestrom P et al. Semliki forest virus produced in the absence of the 6K protein has an altered spike structure as revealed by decreased membrane fusion capacity. Virology 2004; 325:200-6.
53. Yao JS, Strauss EG, Strauss JH. Interactions between PE2, E1 and 6K required for assembly of alphaviruses studied with chimeric viruses. J Virol 1996; 70:7910-20.
54. Lusa S, Garoff H, Liljestrom P. Fate of the 6K membrane protein of Semliki forest virus during virus assembly. Virology 1991; 185:843-6.

55. Lemberg MK, Martoglio B. Requirements for signal peptide peptidase-catalyzed intramembrane proteolysis. Mol Cell 2002; 10:735-44.
56. Weihofen A, Binns K, Lemberg MK et al. Identification of signal peptide peptidase, a presenilin-type aspartic protease. Science 2002; 296:2215-8.
57. Weihofen A, Martoglio B. Intramembrane-cleaving proteases: controlled liberation of proteins and bioactive peptides. Trends Cell Biol 2003; 13:71-8.
58. Wolfe MS, Kopan R. Intramembrane proteolysis: theme and variations. Science 2004; 305:1119-23.
59. Friedmann E, Lemberg MK, Weihofen A et al. Consensus analysis of signal peptide peptidase and homologous human aspartic proteases reveals opposite topology of catalytic domains compared with presenilins. J Biol Chem 2004; 279:50790-8.
60. Friedmann E, Hauben E, Maylandt K et al. SPPL2a and SPPL2b promote intramembrane proteolysis of TNFalpha in activated dendritic cells to trigger IL-12 production. Nat Cell Biol 2006; 8:843-8.
61. Krawitz P, Haffner C, Fluhrer R et al. Differential localization and identification of a critical aspartate suggest nonredundant proteolytic functions of the presenilin homologues SPPL2b and SPPL3. J Biol Chem 2005; 280:39515-23.
62. Fluhrer R, Haass C. Signal peptide peptidases and gamma-secretase: cousins of the same protease family? Neurodegener Dis 2007; 4:112-6.
63. Lemberg MK, Martoglio B. On the mechanism of SPP-catalysed intramembrane proteolysis; conformational control of peptide bond hydrolysis in the plane of the membrane. FEBS Lett 2004; 564:213-8.
64. Srinivas RV, Bernstein H, Oliver C et al. Calmodulin antagonists inhibit human immunodeficiency virus-induced cell fusion but not virus replication. AIDS Res Hum Retroviruses 1994; 10:1489-96.
65. Bryceson YT, Long EO. Line of attack: NK cell specificity and integration of signals. Curr Opin Immunol 2008; 20:344-52.
66. Braud V, Jones EY, McMichael A. The human major histocompatibility complex class Ib molecule HLA-E binds signal sequence-derived peptides with primary anchor residues at positions 2 and 9. Eur J Immunol 1997; 27:1164-9.
67. Bland FA, Lemberg MK, McMichael AJ et al. Requirement of the proteasome for the trimming of signal peptide-derived epitopes presented by the nonclassical major histocompatibility complex class I molecule HLA-E. J Biol Chem 2003; 278:33747-52.
68. Braud VM, Tomasec P, Wilkinson GW. Viral evasion of natural killer cells during human cytomegalovirus infection. Curr Top Microbiol Immunol 2002; 269:117-29.
69. Tomasec P, Braud VM, Rickards C et al. Surface expression of HLA-E, an inhibitor of natural killer cells, enhanced by human cytomegalovirus gpUL40. Science 2000; 287:1031.
70. Ulbrecht M, Martinozzi S, Grzeschik M et al. Cutting edge: the human cytomegalovirus UL40 gene product contains a ligand for HLA-E and prevents NK cell-mediated lysis. J Immunol 2000; 164:5019-22.
71. Lavie M, Goffard A, Dubuisson J. Assembly of a functional HCV glycoprotein heterodimer. Curr Issues Mol Biol 2007; 9:71-86.
72. McLauchlan J, Lemberg MK, Hope G et al. Intramembrane proteolysis promotes trafficking of hepatitis C virus core protein to lipid droplets. EMBO J 2002; 21:3980-8.
73. Hussy P, Langen H, Mous J et al. Hepatitis C virus core protein: carboxy-terminal boundaries of two processed species suggest cleavage by a signal peptide peptidase. Virology 1996; 224:93-104.
74. Okamoto K, Moriishi K, Miyamura T et al. Intramembrane proteolysis and endoplasmic reticulum retention of hepatitis C virus core protein. J Virol 2004; 78:6370-80.
75. Randall G, Panis M, Cooper JD et al. Cellular cofactors affecting hepatitis C virus infection and replication. Proc Natl Acad Sci USA 2007; 104:12884-9.
76. Okamoto K, Mori Y, Komoda Y et al. Intramembrane processing by signal peptide peptidase regulates the membrane localization of hepatitis C virus core protein and viral propagation. J Virol 2008; Sep;82(17):8349-61. Epub 2008.
77. Targett-Adams P, Hope G, Boulant S et al. Maturation of hepatitis C virus core protein by signal peptide peptidase is required for virus production. J Biol Chem 2008; 283:16850-9.
78. Op De Beeck A, Dubuisson J. Topology of hepatitis C virus envelope glycoproteins. Rev Med Virol 2003; 13:233-41.
79. Cocquerel L, Op de Beeck A, Lambot M et al. Topological changes in the transmembrane domains of hepatitis C virus envelope glycoproteins. EMBO J 2002; 21:2893-902.
80. Griffin SD, Beales LP, Clarke DS et al. The p7 protein of hepatitis C virus forms an ion channel that is blocked by the antiviral drug, Amantadine. FEBS Lett 2003; 535:34-8.
81. Clarke D, Griffin S, Beales L et al. Evidence for the formation of a heptameric ion channel complex by the hepatitis C virus p7 protein in vitro. J Biol Chem 2006; 281:37057-68.
82. Crawshaw SG, Martoglio B, Meacock SL et al. A misassembled transmembrane domain of a polytopic protein associates with signal peptide peptidase. Biochem J 2004; 384:9-17.

83. Loureiro J, Ploegh HL. Antigen presentation and the ubiquitin-proteasome system in host-pathogen interactions. Adv Immunol 2006; 92:225-305.
84. Mayer TU, Braun T, Jentsch S. Role of the proteasome in membrane extraction of a short-lived ER-transmembrane protein. EMBO J 1998; 17:3251-7.
85. Jentsch S, Rumpf S. Cdc48 (p97): a "molecular gearbox" in the ubiquitin pathway? Trends Biochem Sci 2007; 32:6-11.
86. Rumpf S, Jentsch S. Functional division of substrate processing cofactors of the ubiquitin-selective Cdc48 chaperone. Mol Cell 2006; 21:261-9.
87. Richly H, Rape M, Braun S et al. A series of ubiquitin binding factors connects CDC48/p97 to substrate multiubiquitylation and proteasomal targeting. Cell 2005; 120:73-84.
88. Cunha FM, Berti DA, Ferreira ZS et al. Intracellular peptides as natural regulators of cell signaling. J Biol Chem 2008; Sep 5;283(36):24448-59. Epub 2008
89. Heimann M, Roman-Sosa G, Martoglio B et al. Core protein of pestiviruses is processed at the C terminus by signal peptide peptidase. J Virol 2006; 80:1915-21.
90. Henderson RA, Michel H, Sakaguchi K et al. HLA-A2.1-associated peptides from a mutant cell line: a second pathway of antigen presentation. Science 1992; 255:1264-6.

CHAPTER 2

mRNA Translation on the Endoplasmic Reticulum:

Kinetic Advantages to Protein Synthesis on the ER

Samuel B. Stephens and Christopher V. Nicchitta*

Abstract

The endoplasmic reticulum (ER) is the exclusive site of secretory and integral membrane protein synthesis. The process of compartmentalizing protein synthesis to the ER occurs through a positive selection mechanism, the signal recognition particle (SRP) pathway. Growing evidence indicates the ER membrane may also be a major site of synthesis for cytosolic and nucleoplasmic proteins. This idea is highlighted by the finding of substantial overlap in mRNA composition of free and membrane-bound polyribosomes. How mRNAs encoding proteins that lack signal peptides would be targeted to the ER for translation is not known. In the context of current models, the synthesis of cytosolic proteins on the ER raises a number of intriguing questions regarding the biological function(s) of protein synthesis compartmentalization. Given the overlap in the identities of mRNAs present in the cytosol and ER membrane populations, it is not known if there are selective advantages in targeting particular mRNAs to one compartment versus the other. Recent studies indicate cytosolic and ER-bound ribosomes translate mRNAs with distinct kinetics; ER ribosomes display 3-4-fold higher levels of protein synthesis than soluble ribosomes. Potentially, directing mRNAs to the ER for translation may function to increase the relative protein expression. Additionally, studies of cell stress pathways have revealed a bias in the cell's utilization of the ER for synthesis of key stress response (nucleoplasmic) proteins. Emerging from these data is a broadening role for the ER membrane as a key site for global protein synthesis regulation.

Abbreviations

Aminoacyl-tRNA synthetase (ARS), endoplasmic reticulum (ER), eukaryotic elongation factor 1B (eEF1B), eukaryotic initiation factor (eIF), rough microsome (RM), signal recognition particle (SRP), small interfering RNA (siRNA), small hairpin RNA (shRNA), SRP receptor (SR), unfolded protein response (UPR), untranslated region (UTR).

Introduction

Current models describe the division of protein synthesis between the cytosol and endoplasmic reticulum (ER) membrane as a mechanism to segregate the synthesis of secretory and integral membrane proteins from cytosolic proteins.[1-6] In this model, the destination of the protein product dictates the primary compartment of synthesis. Thus, cytosolic proteins are synthesized on cytosolic ribosomes and proteins destined for the secretory pathway are synthesized on ER membrane-bound ribosomes. The origins of this model begin with the pioneering work of George

*Corresponding Author: Christopher V. Nicchitta—Department of Cell Biology, Duke University Medical Center, Durham, North Carolina 27710, USA. Email: c.nicchitta@cellbio.duke.edu

Protein Transport into the Endoplasmic Reticulum, edited by Richard Zimmermann

Palade who examined the cellular organization of secretory protein synthesis and maturation. By 1960, Palade and others had demonstrated the basic organization of the secretory pathway: the synthesis of secretory proteins on ribonucleoprotein particles (ribosomes) attached to the ER membrane, the transfer of newly synthesized proteins into the ER lumen and their progression/maturation through Golgi bodies to their final intracellular destination in secretory (zymogen) granules.[7] As a logical extension of this work, soluble cytosolic ribosomes were proposed as the primary site of synthesis for nonsecretory proteins of the cytosolic matrix.

Extending from Palade's work, a key question emerged in the field of protein synthesis: How are newly synthesized secretory proteins (or nascent polypeptides) targeted to the ER and transferred across the limiting membrane into the lumenal cavity? In a landmark publication, Günter Blobel proposed the Signal Hypothesis, in which he postulated an amino-terminal sequence present on the nascent polypeptide chain functioned as a signal to target the synthesis and translocation of secretory proteins at the ER membrane.[1] Through a series of elegant experiments using a heterologous in vitro translation system (canine pancreatic rough microsomes, wheat germ ribosome/translation factors and rat pituitary preprolactin mRNA), the signal recognition particle (SRP) pathway was demonstrated to directly couple secretory/membrane protein synthesis with protein translocation across the ER membrane.[1,2,8-11] The molecular details of this pathway were determined as follows: SRP recognizes secretory and integral membrane proteins early in synthesis via interaction with a signal sequence on the growing nascent polypeptide chain.[1,2,4,11] A key component of the SRP-ribosome interaction is the transient arrest of translation elongation, which is promoted through the 9-/14 kDa heterodimeric subunit of SRP.[9,12] Although this activity is not essential for translocation in vitro,[12] it likely functions in vivo to prevent completion of secretory/membrane protein products in the cytosolic compartment, which could potentially misfold and aggregate. ER membrane targeting of the ribosome/mRNA/nascent chain complex is driven by interactions of SRP with the heterodimeric SRP receptor (SR) present on the ER,[8,13,14] which allows precise docking of the ribosome/nascent chain complex with the aqueous protein-conducting channel or translocon (see chapter by Wagner et al).[15-19] SRP-SR binding facilitates release of SRP from the ribosome/mRNA/nascent chain complex allowing translation elongation to resume.[9,13,14,20,21] Through interactions of the signal sequence with the hydrophobic core of the translocon, most notably Sec61,[15,17,22-24] the nascent polypeptide chain is cotranslationally translocated through the protein-conducting channel into the ER lumen.[1,19] As a component of the translocation machinery, ER resident chaperones such as BiP (GRP78) aid in the vectorial transport of nascent chains into the ER lumen via successive rounds of ATP-dependent chaperone-nascent chain binding and release (for details see chapters by Wilkinson & Stirling and Dudek et al, respectively).[25-27] Following termination of protein synthesis, ER-bound ribosomes detach from the membrane and rejoin the common cytosolic pool.[1,4,28]

Physiological Role of SRP

Experimental evidence regarding the signal sequence-SRP pathway has largely accumulated from in vitro studies using reconstituted translation components. Although this system has provided remarkably detailed insight into the identification and function of each component in the pathway, it is not entirely clear to what degree SRP functions in vivo to direct mRNAs to the ER membrane. In Saccharomyces cerevisiae, two essential components of SRP function, Srp54p and scR1 (7S RNA), are nonessential for cell viability following chromosomal deletion of each gene either separately or in tandem.[29,30] Similarly, mammalian tissue culture cells can also be tolerant of marked reductions in the levels of SRP components (SRP54, SRP72, SRP14).[31,32] Despite reduced cell doubling times in yeast and mammalian cultures upon inactivation of SRP, these studies demonstrate that, in vivo, the SRP pathway is dispensable for targeting essential secretory and membrane protein precursors to the ER membrane. We are now presented with an unanticipated question: What compensatory and/or redundant pathways function to maintain biogenesis of essential secretory and membrane proteins in the absence of SRP? Two possibilities can be considered, neither of which are mutually exclusive: (1) Secretory/membrane protein precursors are synthesized by soluble ribosomes and maintained in a translocation-competent state via cytosolic chaperones. Subsequently, these proteins are posttranslationally translocated across the ER membrane. (2) Nascent polypeptide chains are cotranslationally translocated via binding of soluble ribosomes to available ER binding sites independent of SRP. Additionally, ER-bound ribosomes may usurp the targeting reaction via de novo initiation of secretory/membrane protein synthesis.[33,34] Evidence for these two models is discussed below.

SRP Independent Protein Translocation

Chromosomal deletions of SRP54 and SCR1 (7S RNA) in S. cerevisiae (srp54Δ and scR1Δ), as well as overexpression of a dominant negative allele of SRP54 (*SRP54dn*), result in translocation defects, as demonstrated by the accumulation of unprocessed secretory protein precursors.[29,35] Consistent with the posttranslocation hypothesis, these mutations elicit a rapid and marked up-regulation of the Hsp70 cytosolic chaperone family members Hsp70p and SSA1p/SSA2p and the ER resident chaperone Kar2p (BiP/GRP78)[35,36] which facilitate protein folding and translocation at the ER membrane.[25,27,37,38] However, both the translocation defects and upregulation of heat shock chaperones observed in the *SRP54dn* strain are transient (12-16 hrs) suggesting the chaperone response is a protective effect rather than an adaptive response.[35] Given that the Hsp70 family of chaperones are also involved in protein degradation, the early onset chaperone response may function to target mislocalized (translocation defective) secretory/membrane proteins for degradation rather than to assist in posttranslational translocation as previously proposed. Consistent with this, transcriptional profiling of *SRP54dn* cells identified a two-fold or greater upregulation of protein degradation components.[35]

Following the transient chaperone response to overexpression of the dominant negative SRP54 allele (*SRP54dn*), the expression of genes involved in protein biosynthesis (i.e., ribosomal proteins) is significantly reduced.[35] This observation suggests that decreasing the protein synthesis load compensates for inefficient ER targeting of nascent secretory/membrane polypeptides. Indeed, submaximal concentrations of the reversible elongation inhibitor cycloheximide correct the translocation defects observed in the *SRP54dn* strain. Similarly, low doses of cycloheximide or anisomycin normalize expression of secreted proteins in mammalian cells with reduced SRP14 expression.[32] These data suggest that by reducing elongation rates, the time-frame for ribosome-membrane binding is expanded to accommodate a less-efficient, but functionally redundant pathway for targeting secretory/membrane protein synthesis to the ER independent of SRP.

Although experimentally less well-developed than the yeast system, suppression of SRP expression in mammalian tissue culture cells has demonstrated that membrane protein precursors can vary dramatically in their requirements for SRP function.[31] Using an siRNA-mediated suppression screen, knockdown of the 54 kDa or 72 kDa subunits of SRP inhibits apoptosis mediated by the TRAIL death receptor DR-4, but not DR-5 or the Fas receptor. Stable suppression of either SRP54 or SRP72 led to a similar phenotype, indicating that differential rates of receptor turnover did not contribute to these observations. Why DR-4, but not DR-5 requires SRP function for proper plasma membrane insertion remains unclear. Whether DR-4 and DR-5 differ in their requirements for SRP due to differences in their signal peptide and targeting efficiencies or, alternatively, if a component of the secretory maturation pathway specific to DR-4 is SRP dependent and is thus limiting, has not been determined. Nevertheless, in contrast to yeast, posttranslational translocation is not known to occur in mammalian cells. Thus, a pathway supporting cotranslational translocation remains the prominent mechanism for maintaining secretory/membrane protein biogenesis in the absence of a functional SRP pathway. How this occurs is not known. Given current models of protein synthesis segregation,[6] the studies outlined above underscore the importance of identifying additional or redundant mechanisms for targeting secretory/membrane protein synthesis to the ER, which compensate for disruptions in the SRP targeting pathway.

Coupling the Translation Cycle with Ribosome Binding to the ER Membrane

Current models of protein synthesis compartmentalization dictate that ribosome exchange on the ER membrane directly parallels the mRNA translation cycle; initiation and early elongation steps are coupled to ER membrane targeting whereas termination of translation signals ribosome release.[1,3,4,6] In the following sections, we will juxtapose key experimental observations regarding the translation cycle and ribosome exchange on the ER membrane with predictions of current models. For clarification, it is necessary here to define a number of assumptions made in the SRP model regarding protein synthesis compartmentalization as follows: (1) Translation initiation occurs only on soluble, cytosolic ribosomes, which indiscriminately bind and translate mRNAs.[39,40] In this view, mRNA translation in the cytosol is the default pathway. (2) Targeting mRNAs to the ER membrane requires synthesis of a topogenic signal sequence within the nascent polypeptide chain (i.e., protein synthesis dependent) as a positive selection mechanism.[2,8,11] Thus mRNAs encoding nonsignal sequence containing proteins, such as cytosolic and nucleoplasmic proteins, are predicted to be restricted to the soluble cytosol by default, as they lack a positive ER targeting signal. (3) The signal sequence provides an anchor for the ribosome/mRNA complex to the ER membrane, via interactions of the nascent chain with the core component of the translocon, Sec61.[22,23,28,41] (4) Termination of translation and the ensuing release of the nascent polypeptide chain results in ribosome detachment from the ER membrane.[28,40] In this manner, the ribosome cycle on the ER is predicted to be transient, with termination providing the necessary signal for recycling ribosomes to the cytosol for subsequent translation events.

Ribosome Exchange on the ER Membrane

Ribosome binding to the ER membrane is asymmetric with the large 60S subunit providing the necessary membrane contacts as well as the nascent chain exit tunnel.[42,43] Early studies of ribosome-membrane binding investigated the exchange kinetics of isotopically labeled ribosomal subunits with rat liver rough microsomes (RM).[44] Following release of the nascent polypeptide chain, the authors identified distinct kinetics for release of the small and large subunits; the 60S ribosomal subunit was tightly bound to the ER membrane exhibiting very slow exchange kinetics, whereas the 40S subunit could freely exchange between soluble and membrane-bound states. Similar reports in vivo support this observation and indicate that 60S ribosomal subunits are bound to the ER membrane in the absence of ongoing protein synthesis (i.e., following termination).[28,40,45] Indeed it is now well established that high ionic strength conditions (monovalent cations such as 0.5 M KCl) are typically necessary to displace nontranslating 60S/80S ribosomes from RM. In fact, the current model of ribosome binding to the ER membrane indicates that the 60S subunit directly interacts with the core component of the translocon, the Sec61 complex with 5 nM affinity.[23,46-48] Both biochemical and structural evidence have demonstrated the composition of the ribosome-translocon complex is independent of the translation status of the ribosome.[46,48-50] Thus, the full complement of ribosome-membrane contacts can occur in the absence of an anchoring nascent polypeptide chain. These observations indicate that ribosome binding to the ER membrane can be independent of protein synthesis and suggest that regulatory features in addition to termination can influence ribosome release from the ER.

Coupling of Termination with ER Ribosome Release

To investigate the coupling of protein synthesis termination with ribosome release from the ER membrane in vivo, the fate of membrane-bound ribosomes following pharmacological inhibition of translation initiation has been investigated in a cell culture model.[49,51] In this system, inhibition of translation initiation elicits global termination as translating ribosomes continue elongation and terminate, yet are unable to reassemble or recycle to subsequent translation events. The SRP model predicts that posttermination ribosomes will accumulate in the soluble cytosol because termination results in ribosome-membrane detachment;[1,3,4,6,28] however, data described above examining ribosome exchange on the ER predicts the converse, i.e., that stable ribosome binding to the membrane occurs in the absence of translation. In two parallel studies, using either morphometric analysis of electron micrographs from tissue culture cells or velocity sedimentation experiments using biochemically-derived ER fractions, 60S/80S ribosomes were demonstrated to remain stably associated with the ER translocation machinery following translation termination.[49,51] In a complementary set of studies using either activation of the ER stress pathway, the unfolded protein response (UPR), or picornaviral infection to inactivate initiation complex assembly, physiological suppression of initiation resulted in the accumulation and maintenance of ER-bound 80S ribosomes, similar to results obtained with pharmacological agents described above.[52,53] These studies demonstrate that the physiological process of ribosome release from the ER membrane is regulated independently of the termination reaction.

De Novo Initiation by Membrane-Bound Ribosomes

The findings presented above have resurrected a series of fundamental and as yet unanswered questions first proposed by George Palade.[5] If termination and ribosome detachment are not directly coupled, what additional signals are required for ribosome exchange? Is initiation of mRNA translation restricted to the cytosol, with ribosome traffic to the ER occurring as a cotranslational event, or can both ER-bound and soluble ribosomes function in translation initiation? Furthermore, can ER-bound ribosomes indiscriminately select and translate mRNAs irrespective of the final protein product's destination?

To investigate the significance of de novo translation initiation by ER-bound ribosomes, a series of experiments were performed utilizing the well-established in vitro translation system consisting of canine pancreatic RM and rabbit reticulocyte lysate.[54] In these experiments, mild proteolytic digestion of RM, which disables the SRP receptor,[55-57] was used to distinguish SRP-dependent and independent nascent chain targeting/translocation events. Consistent with predictions of the SRP model,[1,3,4] when translation was initiated by soluble ribosomes from reticulocyte lysate, the SRP receptor was essential for efficient cotranslational targeting and translocation of the preprolactin 86-mer into canine microsomes.[54] Conversely, when RM was the only source of ribosomes provided, proteolytic digestion of the SRP receptor did not diminish targeting or translocation of the preprolactin 86-mer. Importantly, these data not only demonstrate the canonical targeting reaction of a soluble ribosome to the ER via SRP, but also document that the direct binding and translation of a message via a membrane-bound ribosome can operate independent of SRP function. This pathway, which directly complements the SRP pathway, may serve as a compensatory mechanism for continued ER targeting and translocation following inactivation of essential SRP components.

It is worth noting here that the components that regulate targeting of the ribosome/nascent chain complex to the ER membrane are biochemically distinct from those regulating translocation.[58,59] The transfer of the nascent polypeptide chain from the SRP-SR complex requires a "second" recognition event of the signal sequence by components of the translocon.[60,61] In support of this function, experiments using soluble ER-derived ribosome-translocon complexes have demonstrated the translocon is sufficient for both signal sequence recognition and nascent chain translocation independent of additional membrane components such as SR.[49] Thus, the inherent recognition function of the translocon in identifying signal sequences and stimulating translocation, which may be inefficient for targeting secretory/membrane protein synthesis to the ER, may nevertheless be sufficient to coordinate the biogenesis of essential secretory and membrane proteins in the absence of a competent SRP pathway.[34]

mRNA Populations of the Cytosol and ER Membrane Compartments

If ER ribosomes indeed possess the capacity for direct binding of mRNAs and subsequent translation initiation without regard to the final destination of the encoded protein, then mRNAs encoding nonsignal sequence proteins (i.e., soluble cytosolic) are predicted to be present on ER ribosomes. To examine this, genome-wide studies have been performed in yeast, fly, rodent and human systems and the mRNA populations of the cytosol and ER membrane compartments compared using kinetic cDNA hybridizations,[62,63] sequenced cDNA libraries[64] and more recently, cDNA microarrays.[65-67] In these studies, the two subcellular fractions were obtained by buoyant density centrifugation, differential centrifugation and/or sequential detergent (digitonin) extraction. As expected, mRNAs encoding secretory and membrane proteins are highly enriched in ER-bound ribosome fractions, consistent with the protein synthesis role of the ER membrane.[7] However, in contrast to predictions from current models of mRNA selection,[6] these studies also revealed a broad representation of mRNAs encoding cytosolic and nucleoplasmic proteins in ER membrane fractions. Moreover, discrete subpopulations of mRNAs such as those encoding the stress response transcription factor XBP-1 (Hac1p in yeast) and the heat shock chaperone Hsp90 were (highly) enriched in ER ribosome fractions. Single-cell in situ hybridizations have confirmed that indeed, the subcellular distributions of mRNAs such as Hsp90 display a noncanonical pattern of ER enrichment.[65] Although the biological function of these events is not yet clear, the selection of mRNAs for translation on the ER membrane cannot be fully explained in current models of mRNA segregation. Here again, these studies underscore the importance of identifying additional or redundant mechanisms for targeting messages and/or translating ribosomes to the ER membrane independent of the SRP pathway.

How Are Ribosomes Recycled on the ER Membrane?

The observations described above highlight three important inconsistencies within current models of protein synthesis compartmentalization. (1) The obligate pathway for ER ribosomes terminating translation is not release into the cytosol, but rather continued association with the membrane. (2) Translation initiation is not restricted to soluble ribosomes, but may also occur on ER-bound ribosomes. (3) mRNAs can be selected for translation on the ER membrane irrespective of the encoded protein product. With the consideration of these additional pieces of information, we offer an addendum to the current model of ribosome exchange as follows:

As the pre-initiation complex assembles on an mRNA and scans for the AUG start site, a 60S subunit from either the cytosol or ER can be selected for the formation of the mature 80S ribosome. The selection process may be stochastic, perhaps influenced by ribosome subunit availability and/or proximal location of the 48S initiation complex to the ER membrane. Current views describe this process only in the context of a soluble, cytosolic 60S subunit; here we will focus on the selection of a membrane-bound large subunit in an analogous pathway. Following joining of a membrane-bound 60S subunit with the 48S initiation complex, the ER-bound 80S ribosome begins translation. Notably in this model, translation on an ER-bound ribosome occurs prior to the emergence of a topogenic signal sequence, typically viewed as critical for targeting soluble ribosomes synthesizing a secretory/membrane protein to the ER. Thus, mRNAs encoding either secretory/membrane or cytosolic proteins can enter into this pathway. The ability of a membrane-bound ribosome to indiscriminately initiate translation is analogous to the function currently ascribed solely to cytosolic ribosomes.

In the case of an ER-bound ribosome initiating synthesis of a secretory or membrane protein, the recognition function of the translocon is utilized for coupling translation with protein translocation, through direct interaction with the signal peptide on the nascent polypeptide chain. Importantly, this event operates independent of SRP-mediated targeting as the ribosome/mRNA/nascent chain is already present on the ER. Throughout the elongation cycle, the ribosome remains firmly attached to the ER membrane via the anchoring nascent polypeptide chain. Upon termination and disassembly of the 80S ribosome, the large 60S subunit remains tightly associated with the translocon awaiting a new translation cycle.

In the case of an ER-bound ribosome initiating synthesis of a cytosolic protein, the emerging nascent polypeptide must extend laterally along the ER membrane surface as it lacks a positive signal for opening the translocon channel. In this scenario, it is essential that the translocon remains in a translocation-incompetent state to maintain the ion impermeability of the ER. During this elongation cycle, we propose ribosomes detach from the ER membrane. Perhaps distinct ribosome conformations accessed during successive rounds of peptide bond formation and translocation to the downstream codon result in reduced ribosome-translocon affinity. Additionally, the nontranslocated (cytosolic) nascent polypeptide chain may perturb the ribosome-translocon interaction. In the absence of an anchoring nascent chain, these events result in ribosome release from the ER and continued synthesis of the cytosolic protein in the cytosol. Following termination and disassembly of the 80S ribosome, the once membrane-bound ribosomal subunits have re-entered the cytosolic pool for subsequent translation events.

An important consideration in the synthesis of a cytosolic protein on the ER membrane is the apparent paradox concerning the tight junction formed between the ribosome and the protein-conducting channel.[15,17-19] Whereas it is generally accepted that interactions between the nascent chain's signal sequence, the ribosome and the translocon serve as a signal for opening the channel, it is conceivable that the channel would remain in a closed conformation or translocation-incompetent state during synthesis of a cytosolic protein. Similar ideas have been postulated for the synthesis of cytosolic domains of integral membrane proteins, which pose a similar threat to the leakage of ions out of the ER lumen.[68] Recent structural studies using cryo-electron microscopy to visualize native and reconstituted mammalian ribosome/translocon complexes have provided insight into this seeming paradox identifying a large (15-20Å) gap between the ribosome exit tunnel and the translocon.[46-48] Importantly, this gap has been postulated to be of sufficient size to allow lateral egress of intracellular (cytosolic) domains of membrane proteins as well as strictly cytosolic nascent polypeptide chains. Thus, from a structural perspective, synthesis of a cytosolic protein by an ER-bound ribosome is not refractory to maintaining the binding contacts necessary for the ribosome-translocon interaction.

In summary, the pathway described here utilizes the elongation component of the translation cycle as both a positive and a negative selection event for ribosome-membrane exchange. Cytosolic ribosomes synthesizing a secretory/membrane protein are cotranslationally targeted (positive) to the ER via the SRP pathway during elongation; conversely, ER-bound ribosomes synthesizing a cytosolic protein detach (negative) from the ER membrane during the elongation cycle and re-enter the soluble cytosol. Together these events complete the cycle of ribosome-membrane binding and release.

Mechanism(s) of mRNA Selection

From the viewpoint of existing models, the observation that mRNAs encoding cytosolic proteins can be partitioned to and translated on, ER-bound ribosomes is unexpected.[2,6] Nevertheless, rigorous examinations of mRNA partitioning between soluble and membrane-bound ribosomes, conducted in multiple cell types and tissues, in lower and higher eukaryotes and by multiple fractionation techniques, have concluded that the mechanism of mRNA partitioning in the eukaryotic cell is complex.[62-67] Little is known regarding the mechanism(s) for how mRNA partitioning patterns is conferred, but is likely to occur through multiple and perhaps independent pathways. In the preceding section, we have described a process for the general targeting of mRNAs to the ER membrane, using the inherent recognition function (de novo initiation) of a membrane-bound ribosome to begin translation without regard to the encoded protein's identity. In this model, the stability of the ribosome-translocon interaction functions as a selection mechanism to maintain or release translating ribosomes from the ER and thus may contribute to the relative partitioning of mRNAs between the cytosol and ER compartments. In addition to this pathway, the identification of mRNAs such as Hsp90 and XBP-1, which are highly enriched on the ER membrane, suggests that more specific mechanisms also exist for localizing transcripts to the membrane.

Typical mechanisms of mRNA localization are the result of interactions of specific RNA cis-elements, which are usually located in the untranslated regions (UTR), with an RNA binding protein(s) (trans-factors) that actively target an mRNA to a discrete (subcellular) locale(s). The localization of β-actin mRNA to the cell cortex is a prominent example of this mechanism in mammalian cells. The β-actin message is recognized by a 54 nucleotide cis-sequence present in the 3′-UTR and trafficked along microtubules to the cell cortex via interactions with its cognate RNA binding protein, ZBP-1.[69-71] Extending from this concept, the enrichment of specific mRNAs encoding cytosolic/nucleoplasmic proteins on the ER (such as XBP-1 and Hsp90) may occur through direct interactions of an RNA localization cis-element(s) with an ER membrane RNA receptor. This localization could be direct or require individual RNA binding protein(s) specific to these messages to interact with a cognate ER membrane (RNA binding protein) receptor analogous to the β-actin localization pathway. Additionally, it is possible that a given mRNA species may be represented in multiple subpopulations, defined by different cohorts of RNA binding proteins and thus differ in relative partitioning between the cytosol and ER compartments.[72]

Importantly, mRNA-mediated ER localization pathways may not be exclusive to messages encoding cytosolic/nucleoplasmic proteins, rather this can also occur in the context of mRNAs encoding secretory/membrane proteins. The mRNA encoding the abundant ER resident chaperone GRP94 (gp96) was identified as an unanticipated example of such a pathway.[73] Examination of HeLa cells stably expressing an shRNA targeting the SRP subunit, SRP54, demonstrated that GRP94, calreticulin, BiP and TRAPδ mRNAs were exclusively localized to the ER membrane in cells with markedly limiting or absent SRP function, indistinguishable from control (wild type) cultures. Further studies revealed that deletion of the (encoded) signal peptide did not displace GRP94 mRNA from the ER membrane with a concomitant increase in the cytosol, but rather this message continued to be highly enriched on the ER membrane. This observation provides key evidence that mRNAs encoding secretory proteins can and are targeted to the ER membrane independent of SRP.

Emerging Roles for the ER Membrane in Global Protein Synthesis

As described above, the pattern of mRNA partitioning in eukaryotic cells is more complex than previously predicted, perhaps requiring multiple (independent) targeting and selection pathways. For example, random sampling may filter some mRNAs between the cytosol and ER, whereas additional mechanisms such as the SRP pathway or specific mRNA receptors are employed in other cases to traffic mRNAs between the two compartments. A key consideration here is whether mRNAs of either class (secretory/membrane and cytosolic) continually shuttle between the cytosol and ER membrane in dynamic equilibrium, or remain continually associated with one primary compartment following initial deposition. Potentially, environmental cues relevant to a specific cell's (tissue) metabolic demands may influence the mobilization of mRNA populations between the two compartments. From a functional perspective, the role of the ER membrane in synthesizing secretory and membrane proteins is quite clear; in contrast, with regards to proteins residing in the cytosol and nucleoplasm, the role of membrane-bound ribosomes in their synthesis is only now emerging and discussed below.

Kinetic Advantages in ER-Directed mRNA Translation

Soluble and membrane-bound ribosomes are essentially both cytosolic (i.e., present on the cytosolic face of the ER membrane). In this respect, the relative rates of translation and polyribosome assembly are not predicted to differ between the two compartments, assuming equal opportunity and accessibility to necessary translation factors. However during cellular stress responses that reduce translation initiation, clear differences have been identified between the cytosol and ER compartments with regards to polyribosome assembly.[52,53] Using either picornavirus infection to suppress cap-dependent mRNA translation or ER stress (UPR) to inhibit eIF2 activity, polyribosomes are spatially restricted to the ER membrane with complete disassembly of polyribosomes into free mRNA and ribosome occurring in the cytosol. As a result, mRNAs that were present on both cytosolic and ER-bound ribosomes under homeostatic conditions display ER-restricted loading into polyribosomes under conditions of limiting initiation factor availability. Two mutually exclusive conclusions are suggested from these studies: (1) Soluble and membrane bound ribosomes continue to function equivalently in protein synthesis, yet display distinct stabilities and kinetics of polyribosome assembly. (2) Translation on soluble vs membrane-bound ribosomes is kinetically distinct such that reducing translation initiation rates (cell stress) unveils a divergence in polyribosome assembly.

To examine these possibilities, a kinetic evaluation of soluble and membrane bound ribosome translation was performed in a cell culture model.[74] Using a pulse-chase procedure, initiation/elongation (pulse) and elongation/termination (pulse-chase) rates of translation were monitored in vivo. Remarkably, under homeostatic growth conditions, mRNA translation on soluble vs membrane-bound ribosomes operates with distinct kinetics; ER-bound ribosomes exhibit 3-4-fold higher rates of protein synthesis as compared with cytosolic ribosomes. Moreover, despite dramatic differences in the relative synthesis rates of the two populations, the total amounts of newly synthesized proteins entering either the cytosol or secretory/membrane compartments are equivalent. These findings suggest that the ER membrane, in addition to the biogenesis of secretory/membrane proteins, also contributes a substantial quantity of newly synthesized proteins to the cytosolic pool. In this respect, the ER membrane may be regarded as the primary site of synthesis for secretory/membrane proteins and a major contributor to the synthesis of cytosolic proteins as well. In addition, these data demonstrate that the partitioning of an mRNA to the cytosol or ER membrane can have dramatic consequences on its translation rate.

Coupling mRNA Translation with tRNA Channeling on the ER Membrane

A number of observations have suggested that the processes of mRNA translation and tRNA aminoacylation/deacylation are directly coupled to facilitate or enhance translation through a pathway termed tRNA channeling.[76,77] In this pathway, newly charged tRNA is vectorially transferred from the aminoacyl-tRNA synthetase (ARS) to initiation/elongation factor complexes for direct use in mRNA translation.[78-83] Additionally, tRNAs can be channeled through iterative cycles of protein synthesis without exchange into a soluble pool.[76,77] The proposed coupling of the tRNA and translation cycles would concentrate translation components at sites of utilization, to increase the efficiency of the translation reaction.

In addition to the enhanced rates of translation observed for ER-bound ribosomes, distinct kinetics for tRNA turnover between the two compartments were also demonstrated in the context of the elongation inhibitors, cycloheximide and emetine.[74] In the cytosol, tRNAs are rapidly recycled (deacylated and re-acylated) irrespective of protein synthesis rates, likely occurring through active tRNA hydrolase activity; in contrast, tRNA turnover on the ER is tightly coupled to the translation status of ER ribosomes. It is intriguing to envision this observation as a manifestation of an ER-localized tRNA channeling process, which may contribute to the enhanced rates of translation on membrane-bound ribosomes.

Two of the critical components regulating tRNA aminoacylation and subsequent elongation factor binding have been identified on the ER membrane: the multi-enzyme aminoacyl-tRNA synthetase complex, which contains 9 of the 20 cellular ARSs[84,85] and eukaryotic elongation factor 1B (eEF1B),[86] which is the guanine nucleotide exchange factor for eEF1A. Although the mechanism(s) governing assembly of the multi-enzyme ARS complex on the ER membrane has not been identified, the association of eEF1B with the ER membrane occurs through interactions of the delta subunit with the ER integral membrane protein, kinectin.[87] In a recent publication, the subcellular localization of eEF1B was sufficient to preferentially enhance protein synthesis in one compartment with a corresponding decrease in the other compartment.[88] In the case of differential sensitivity of tRNA turnover to protein synthesis suppression,[74] perhaps, this reflects a preference for tRNAs on the ER to be present in initiation/elongation factor complexes and thereby protected from rapid hydrolase-mediated deacylation. Future studies are necessary to further dissect the biological significance of this pathway and ramifications for protein synthesis regulation.

Future Perspective: Expanding the Role of the ER Membrane

Historically, the primary role of the rough ER has been the biogenesis of secretory and integral membrane proteins. The selection and targeting of translating ribosomes to the ER occurs largely through the SRP pathway. However, accumulating evidence, both genetic and biochemical, indicate additional pathways can also function in this process. A likely mechanism proposed here is the de novo initiation by membrane-bound ribosomes. In this pathway, ER-bound ribosomes do not detach from the membrane following translation termination, but rather remain in a translation competent state. Analogous to soluble ribosomes, ER-bound ribosomes initiate translation of an mRNA irrespective of the encoded protein's final destination (cytosolic or secretory/membrane). At this point in the targeting cycle, the emergence of a signal sequence on a secretory/membrane protein is both necessary and sufficient for opening the translocon and subsequent translocation independent of SRP components. In the event of an ER-bound ribosome initiating synthesis of a cytosolic protein (i.e., lacking a signal sequence), the translocon would remain closed, thereby preventing inappropriate entry into the ER lumen. How the stability of the ribosome-translocon interaction is affected in this context remains speculative, but we predict that in the absence of an anchoring nascent polypeptide chain, translating ribosomes would detach from the membrane and re-enter the cytosolic pool. Additionally, specific sequences in the nascent polypeptide chain may exert either positive or negative effects on ribosome-membrane binding and thereby contribute to relative dynamics of ribosome/mRNA-membrane exchange.

With the finding that mRNAs encoding cytosolic and nucleoplasmic proteins are present in ER polyribosomes, the role of the ER in protein synthesis is expanding. In addition to providing the entry point for the secretory pathway, the ER also serves as a platform for the synthesis of cytosolic proteins. As a direct consequence of the ER's robust translation activity, we anticipate ER ribosomes provide a significant contribution to the synthesis of the cytosolic protein population. With regards to the ER's enhanced translation activity, tRNA channeling is proposed as a key kinetic parameter, however, additional translation component assembly, such as initiation factor complexes, may also contribute to this process. It is worth speculating that different cell types (tissues) may exhibit distinct patterns of noncanonical mRNAs populating the ER membrane based in part, upon the precise metabolic demands of the cell (and/or tissue). By utilizing the distinct translation activities of the cytosol and ER, the cell may exercise exquisite control over the level of synthesis of a given protein through regulating the relative subcellular distribution of the mRNA.

Acknowledgements

The authors would like to thank Drs. Rebecca Dodd, Rachel Lerner and Brook Pyhtila for their insightful comments and suggestions. This work was supported by a National Institute of Health grant to C.V.N (GM077382) and an American Heart Association Predoctoral Fellowship to S.B.S (0515333U).

References

1. Blobel G, Dobberstein B. Transfer of proteins across membranes. I. Presence of proteolytically processed and unprocessed nascent immunoglobulin light chains on membrane-bound ribosomes of murine myeloma. J Cell Biol 1975; 67:835-851.
2. Blobel G, Dobberstein B. Transfer of proteins across membranes. II. Reconstitution of functional rough microsomes from heterologous components. J Cell Biol 1975; 67:852-862.
3. Blobel G, Walter P, Chang CN et al. Translocation of proteins across membranes: the signal hypothesis and beyond. Symp Soc Exp Biol 1979; 33:9-36.
4. Lingappa VR, Blobel G. Early events in the biosynthesis of secretory and membrane proteins: the signal hypothesis. Recent Prog Horm Res 1980; 36:451-475.
5. Palade G. Intracellular aspects of the process of protein synthesis. Science 1975; 189:347-358.
6. Alberts B, Johnson A, Lewis J et al. Molecular biology of the Cell. 4th ed. New York, NY: Garland Science 2002.
7. Siekevitz P, Palade GE. A cytochemical study on the pancreas of the guinea pig 5. In vivo incorporation of leucine-1-C14 into the chymotrypsinogen of various cell fractions. J Biophys Biochem Cytol 1960; 7:619-630.
8. Walter P, Blobel G. Translocation of proteins across the endoplasmic reticulum. II. Signal recognition protein (SRP) mediates the selective binding to microsomal membranes of in-vitro-assembled polysomes synthesizing secretory protein. J Cell Biol 1981; 91:551-556.
9. Walter P, Blobel G. Translocation of proteins across the endoplasmic reticulum III. Signal recognition protein (SRP) causes signal sequence-dependent and site- specific arrest of chain elongation that is released by microsomal membranes. J Cell Biol 1981; 91:557-561.
10. Walter P, Blobel G. Purification of a membrane-associated protein complex required for protein translocation across the endoplasmic reticulum. Proc Natl Acad Sci USA 1980; 77:7112-7116.

11. Walter P, Ibrahimi I, Blobel G. Translocation of proteins across the endoplasmic reticulum. I. Signal recognition protein (SRP) binds to in-vitro-assembled polysomes synthesizing secretory protein. J Cell Biol 1981; 91:545-550.
12. Siegel V, Walter P. Elongation arrest is not a prerequisite for secretory protein translocation across the microsomal membrane. J Cell Biol 1985; 100:1913-1921.
13. Gilmore R, Blobel G, Walter P. Protein translocation across the endoplasmic reticulum. I. Detection in the microsomal membrane of a receptor for the signal recognition particle. J Cell Biol 1982; 95:463-469.
14. Gilmore R, Walter P, Blobel G. Protein translocation across the endoplasmic reticulum. II. Isolation and characterization of the signal recognition particle receptor. J Cell Biol 1982; 95:470-477.
15. Crowley KS, Reinhart GD, Johnson AE. The signal sequence moves through a ribosomal tunnel into a noncytoplasmic aqueous environment at the ER membrane early in translocation. Cell 1993; 73:1101-1115.
16. Walter P, Lingappa VR. Mechanism of protein translocation across the endoplasmic reticulum membrane. Annu Rev Cell Biol 1986; 2:499-516.
17. Crowley KS, Liao S, Worrell VE et al. Secretory proteins move through the endoplasmic reticulum membrane via an aqueous, gated pore. Cell 1994; 78:461-471.
18. Simon SM, Blobel G. A protein-conducting channel in the endoplasmic reticulum. Cell 1991; 65:371-380.
19. Gilmore R, Blobel G. Translocation of secretory proteins across the microsomal membrane occurs through an environment accessible to aqueous perturbants. Cell 1985; 42:497-505.
20. Gilmore R, Blobel G. Transient involvement of signal recognition particle and its receptor in the microsomal membrane prior to protein translocation. Cell 1983; 35:677-685.
21. Miller JD, Wilhelm H, Gierasch L et al. GTP binding and hydrolysis by the signal recognition particle during initiation of protein translocation. Nature 1993; 366:351-354.
22. Gorlich D, Prehn S, Hartmann E et al. A mammalian homolog of SEC61p and SECYp is associated with ribosomes and nascent polypeptides during translocation. Cell 1992; 71:489-503.
23. Kalies KU, Gorlich D, Rapoport TA. Binding of ribosomes to the rough endoplasmic reticulum mediated by the Sec61p-complex. J Cell Biol 1994; 126:925-934.
24. Mothes W, Prehn S, Rapoport TA. Systematic probing of the environment of a translocating secretory protein during translocation through the ER membrane. EMBO J 1994; 13:3973-3982.
25. Sanders SL, Whitfield KM, Vogel JP et al. Sec61p and BiP directly facilitate polypeptide translocation into the ER. Cell 1992; 69:353-365.
26. Nicchitta CV, Blobel G. Lumenal proteins of the mammalian endoplasmic reticulum are required to complete protein translocation. Cell 1993; 73:989-998.
27. Vogel JP, Misra LM, Rose MD. Loss of BiP/GRP78 function blocks translocation of secretory proteins in yeast. J Cell Biol 1990; 110:1885-1895.
28. Mechler B, Vassalli P. Membrane-bound ribosomes of myeloma cells. III. The role of the messenger RNA and the nascent polypeptide chain in the binding of ribosomes to membranes. J Cell Biol 1975; 67:25-37.
29. Hann BC, Walter P. The signal recognition particle in S. cerevisiae. Cell 1991; 67:131-144.
30. Felici F, Cesareni G, Hughes JM. The most abundant small cytoplasmic RNA of saccharomyces cerevisiae has an important function required for normal cell growth. Mol Cell Biol 1989; 9:3260-3268.
31. Ren YG, Wagner KW, Knee DA et al. Differential regulation of the TRAIL death receptors DR4 and DR5 by the signal recognition particle. Mol Biol Cell 2004; 15:5064-5074.
32. Lakkaraju AK, Mary C, Scherrer A et al. SRP keeps polypeptides translocation-competent by slowing translation to match limiting ER-targeting sites. Cell 2008; 133:440-451.
33. Potter MD, Nicchitta CV. Ribosome-independent regulation of translocon composition and sec61alpha conformation. J Biol Chem 2000; 275:2037-2045.
34. Potter MD, Seiser RM, Nicchitta CV. Ribosome exchange revisited: a mechanism for translation-coupled ribosome detachment from the ER membrane. Trends Cell Biol 2001; 11:112-115.
35. Mutka SC, Walter P. Multifaceted physiological response allows yeast to adapt to the loss of the signal recognition particle-dependent protein-targeting pathway. Mol Biol Cell 2001; 12:577-588.
36. Arnold CE, Wittrup KD. The stress response to loss of signal recognition particle function in saccharomyces cerevisiae. J Biol Chem 1994; 269:30412-30418.
37. Chirico WJ, Waters MG, Blobel G. 70K heat shock related proteins stimulate protein translocation into microsomes. Nature 1988; 332:805-810.
38. Deshaies RJ, Koch BD, Werner-Washburne M et al. A subfamily of stress proteins facilitates translocation of secretory and mitochondrial precursor polypeptides. Nature 1988; 332:800-805.
39. Mechler B. Membrane-bound ribosomes of myeloma cells. VI. Initiation of immunoglobulin mRNA translation occurs on free ribosomes. J Cell Biol 1981; 88:42-50.

40. Mechler B, Vassalli P. Membrane-bound ribosomes of myeloma cells. II. Kinetic studies on the entry of newly made ribosomal subunits into the free and the membrane-bound ribosomal particles. J Cell Biol 1975; 67:16-24.
41. Adelman MR, Sabatini DD, Blobel G. Ribosome-membrane interaction. Nondestructive disassembly of rat liver rough microsomes into ribosomal and membranous components. J Cell Biol 1973; 56:206-229.
42. Sabatini DD, Tashiro Y, Palade GE. On the attachment of ribosomes to microsomal membranes. J Mol Biol 1966; 19:503-524.
43. Unwin PN. Three-dimensional model of membrane-bound ribosomes obtained by electron microscopy. Nature 1977; 269:118-122.
44. Borgese D, Blobel G, Sabatini DD. In vitro exchange of ribosomal subunits between free and membrane-bound ribosomes. J Mol Bio 1973; 74:415-438.
45. Baglioni C, Bleiberg I, Zauderer M. Assembly of membrane-bound polyribosomes. Nat New Biol 1971; 232:8-12.
46. Beckmann R, Spahn CM, Eswar N et al. Architecture of the protein-conducting channel associated with the translating 80S ribosome. Cell 2001; 107:361-372.
47. Ménétret JF, Neuhof A, Morgan DG et al. The structure of ribosome-channel complexes. Molecular Cell 2000; 6:1219-1232.
48. Morgan DG, Ménétret JF, Neuhof A et al. Structure of the mammalian ribosome–channel complex at 17 Å resolution. J Mol Biol 2002; 324:871-886.
49. Potter MD, Nicchitta CV. Endoplasmic reticulum-bound ribosomes reside in stable Association with the translocon following termination of protein synthesis. J Biol Chem 2002; 277:23314-23320.
50. Snapp EL, Reinhart GA, Bogert BA et al. The organization of engaged and quiescent translocons in the endoplasmic reticulum of mammalian cells. J Cell Biol 2004; 164:997-1007.
51. Seiser RM, Nicchitta CV. The fate of membrane-bound ribosomes following the termination of protein synthesis. J Biol Chem 2000; 275:33820-33827.
52. Lerner RS, Nicchitta CV. mRNA translation is compartmentalized to the endoplasmic reticulum following physiological inhibition of cap-dependent translation. RNA 2006; 12:775-789.
53. Stephens SB, Dodd RD, Brewer JW et al. Stable ribosome binding to the endoplasmic reticulum enables compartment-specific regulation of mRNA translation. Mol Biol Cell 2005; 16:5819-5831.
54. Potter MD, Nicchitta CV. Regulation of ribosome detachment from the mammalian endoplasmic reticulum membrane. J Biol Chem 2000; 275:33828-33835.
55. Meyer DI, Dobberstein B. Identification and characterization of a membrane component essential for the translocation of nascent proteins across the membrane of the endoplasmic reticulum. J Cell Biol 1980; 87:503-508.
56. Meyer DI, Dobberstein B. A membrane component essential for vectorial translocation of nascent proteins across the endoplasmic reticulum: requirements for its extraction and reassociation with the membrane. J Cell Biol 1980; 87:498-502.
57. Walter P, Jackson RC, Marcus MM et al. Tryptic dissection and reconstitution of translocation activity for nascent presecretory proteins across microsomal membranes. Proc Natl Acad Sci USA 1979; 76:1795-1799.
58. Nicchitta C, Migliaccio G, Blobel G. Reconstitution of secretory protein translocation from detergent-solubilized rough microsomes. Methods Cell Biol 1991; 34:263-285.
59. Nicchitta CV, Blobel G. Nascent secretory chain binding and translocation are distinct processes: differentiation by chemical alkylation. J Cell Biol 1989; 108:789-795.
60. Siegel V. A second signal recognition event required for translocation into the endoplasmic reticulum. Cell 1995; 82:167-170.
61. Jungnickel B, Rapoport TA. A posttargeting signal sequence recognition event in the endoplasmic reticulum membrane. Cell 1995; 82:261-270.
62. Mueckler MM, Pitot HC. Structure and function of rat liver polysome populations. I. Complexity, frequency distribution and degree of uniqueness of free and membrane-bound polysomal polyadenylate-containing RNA populations. J Cell Biol 1981; 90:495-506.
63. Mechler B, Rabbitts TH. Membrane-bound ribosomes of myeloma cells. IV. mRNA complexity of free and membrane-bound polysomes. J Cell Biol 1981; 88:29-36.
64. Kopczynski CC, Noordermeer JN, Serano TL et al. A high throughput screen to identify secreted and transmembrane proteins involved in drosophila embryogenesis. Proc Natl Acad Sci USA 1998; 95:9973-9978.
65. Lerner RS, Seiser RM, Zheng T et al. Partitioning and translation of mRNAs encoding soluble proteins on membrane-bound ribosomes. RNA 2003; 9:1123-1137.
66. Diehn M, Bhattacharya R, Botstein D et al. Genome-scale identification of membrane-associated human mRNAs. PLoS Genet 2006; 2:e11.

67. Diehn M, Eisen MB, Botstein D et al. Large-scale identification of secreted and membrane-associated gene products using DNA microarrays. Nat Genet 2000; 25:58-62.
68. Liao S, Lin J, Do H et al. Both lumenal and cytosolic gating of the aqueous ER translocon pore are regulated from inside the ribosome during membrane protein integration. Cell 1997; 90:31-41.
69. Ross AF, Oleynikov Y, Kislauskis EH et al. Characterization of a beta-actin mRNA zipcode-binding protein. Mol Cell Biol 1997; 17:2158-2165.
70. Gu W, Pan F, Zhang H et al. A predominantly nuclear protein affecting cytoplasmic localization of beta-actin mRNA in fibroblasts and neurons. J Cell Biol 2002; 156:41-51.
71. Rodriguez AJ, Shenoy SM, Singer RH et al. Visualization of mRNA translation in living cells. J Cell Biol 2006; 175:67-76.
72. Keene JD, Lager PJ. Post-transcriptional operons and regulons co-ordinating gene expression. Chromosome Res 2005; 13:327-337.
73. Pyhtila B, Zheng T, Lager PJ et al. Signal sequence- and translation-independent mRNA localization to the endoplasmic reticulum. RNA 2008; 14:445-453.
74. Stephens SB, Nicchitta CV. Divergent regulation of protein synthesis in the cytosol and endoplasmic reticulum compartments of mammalian cells. Mol Biol Cell 2008; 19:623-632.
75. Stephens SB, Dodd RD, Lerner RS et al. Analysis of mRNA partitioning between the cytosol and endoplasmic reticulum compartments of mammalian cells. Methods Mol Biol 2008; 419:197-214.
76. Negrutskii BS, Deutscher MP. Channeling of aminoacyl-tRNA for protein synthesis in vivo. Proc Natl Acad Sci USA 1991; 88:4991-4995.
77. Stapulionis R, Deutscher MP. A channeled tRNA cycle during mammalian protein synthesis. Proc Natl Acad Sci USA 1995; 92:7158-7161.
78. Reed VS, Wastney ME, Yang DC. Mechanisms of the transfer of aminoacyl-tRNA from aminoacyl-tRNA synthetase to the elongation factor 1 alpha. J Biol Chem 1994; 269:32932-32936.
79. Negrutskii BS, Shalak VF, Kerjan P et al. Functional interaction of mammalian valyl-tRNA synthetase with elongation factor EF-1alpha in the complex with EF-1H. J Biol Chem 1999; 274:4545-4550.
80. Bec G, Kerjan P, Waller JP. Reconstitution in vitro of the valyl-tRNA synthetase-elongation factor (EF) 1 beta gamma delta complex. Essential roles of the NH2-terminal extension of valyl-tRNA synthetase and of the EF-1 delta subunit in complex formation. J Biol Chem 1994; 269:2086-2092.
81. Bec G, Kerjan P, Zha XD et al. Valyl-tRNA synthetase from rabbit liver. I. Purification as a heterotypic complex in association with elongation factor 1. J Biol Chem 1989; 264:21131-21137.
82. Brandsma M, Kerjan P, Dijk J et al. Valyl-tRNA synthetase from artemia. Purification and association with elongation factor 1. Eur J Biochem 1995; 233:277-282.
83. Negrutskii BS, Budkevich TV, Shalak VF et al. Rabbit translation elongation factor 1 alpha stimulates the activity of homologous aminoacyl-tRNA synthetase. FEBS Lett 1996; 382:18-20.
84. Dang CV, Yang DC, Pollard TD. Association of methionyl-tRNA synthetase with detergent-insoluble components of the rough endoplasmic reticulum. J Cell Biol 1983; 96:1138-1147.
85. Johnson DL, Yang DC. Stoichiometry and composition of an aminoacyl-tRNA synthetase complex from rat liver. Proc Natl Acad Sci USA 1981; 78:4059-4062.
86. Sanders J, Brandsma M, Janssen GM et al. Immunofluorescence studies of human fibroblasts demonstrate the presence of the complex of elongation factor-1 beta gamma delta in the endoplasmic reticulum. J Cell Sci 1996; 109:1113-1117.
87. Ong LL, Er CP, Ho A et al. Kinectin anchors the translation elongation factor-1 delta to the endoplasmic reticulum. J Biol Chem 2003; 278:32115-32123.
88. Ong LL, Lin PC, Zhang X et al. Kinectin-dependent assembly of translation elongation factor-1 complex on endoplasmic reticulum regulates protein synthesis. J Biol Chem 2006; 281:33621-33634.

CHAPTER 3

Translocon Organization in Cells

Deborah E. Aronson and Erik L. Snapp*

Abstract

A protein's function is a product of its inherent structural and biochemical characteristics. Equally critical to protein function are elements of its cellular environment including localization, organization and dynamics, which regulate the capacity of a protein to encounter substrates in cells. The majority of Sec61 translocon research has exploited biochemical approaches to successfully characterize the structure and function of several integral membrane translocon components. However, ensemble biochemical measurements and structural approaches neither provide spatial nor temporal information regarding the environmental context of translocon components. In addition, nonphysiological conditions required for biochemical analyses, such as detergent extraction of membrane proteins, may disrupt weak interactions between transmembrane domains, whereas conversely, cross-linking agents may introduce nonphysiological protein-protein interactions. Here we will discuss how biophysical microscopy techniques have provided new insights into translocon organization in cells and highlight developments in imaging applications, which we anticipate will open new lines of inquiry.

Introduction

The endoplasmic reticulum (ER) is frequently the largest organelle in eukaryotic cells and is able to adopt numerous architectures ranging from a network of branching tubules to stacked cisternae or dense clusters of tubules (Fig. 1).[1-3] These diverse structures can be classified into rough and smooth ER (RER and SER, respectively), as determined by the presence (RER) or absence (SER) of membrane-bound cytoplasmic ribosomes (Fig. 1). Ribosomes on RER are found on stacked arrays of short cisternae and to a lesser extent on tubules (Fig. 2).[2] In contrast, SER exists primarily as ribosome-free densely clustered tubules[2] and highly organized crystalloid stacks.[1,4] How these subdomains form and segregate within an otherwise continuous organelle[5] remains poorly understood.[6] The principles that regulate ribosome binding and density on ER membranes are central to understanding ER subdomain formation and secretory protein synthesis.[1,3,7,8]

Ribosomes associate with the ER predominantly via the Sec61 translocon.[9] An additional factor, the mammalian protein p180, has been shown to stimulate RER proliferation,[10,11] though the significance of p180 in regulation of RER biogenesis remains poorly understood.[9] The mechanisms regulating Sec61 levels, distribution and mobility in cells are all likely to be critical for RER biogenesis. A variety of approaches, including live cell imaging methods, have been employed recently to investigate these mechanisms.

The most basic problem of RER biogenesis is to determine whether ribosomes can bind anywhere on the ER with equal probability. Are Sec61 translocon core components equally available throughout the ER? Using quantitative proteomics, Gilchrist et al determined the relative distributions of Sec61 translocon components in purified rough and smooth ER microsomes.[12] Data in Table 1 reveal that Sec61α in rat liver homogenate, is enriched at least seven-fold in RER compared

*Corresponding Author: Erik L. Snapp—Department of Anatomy and Structural Biology, Albert Einstein College of Medicine, Bronx, NY 10461, USA. Email: esnapp@aecom.yu.edu

Protein Transport into the Endoplasmic Reticulum, edited by Richard Zimmermann

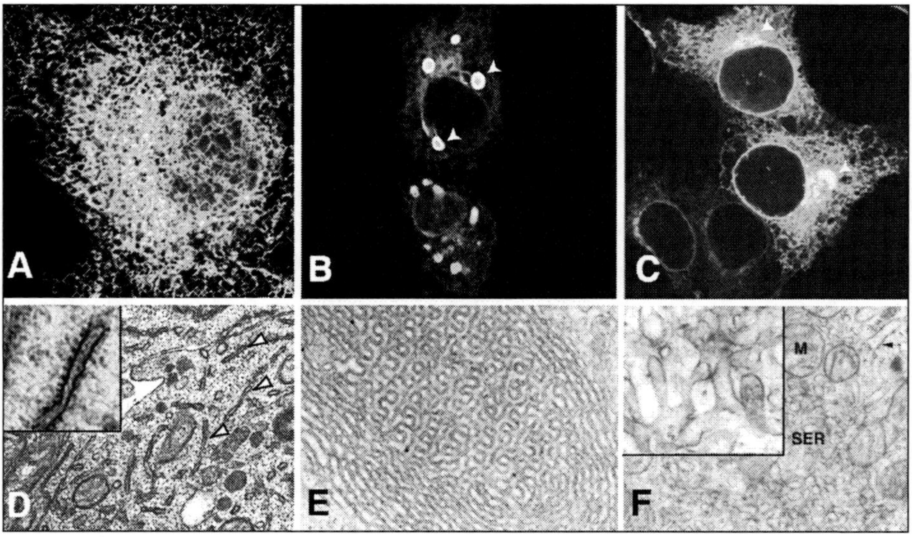

Figure 1. Structures of the endoplasmic reticulum. A-C) Confocal fluorescence micrographs of Cos-7 cells expressing GFP-tagged constructs localized to the ER. A) Branching ER tubular network. B) OSER-containing cells. C) Accumulations of random tubular SER. D-F) Transmission Electron Micrographs of ER structures. D) AR42J cells stimulated to proliferate RER with DMSO for 72 h. White arrows point to ribosome studded cisternae. Inset features magnified RER cistern covered with ribosomes (dense black dots). TEM images were generously provided by Dr. Rachid Sougrat. E) Cos-7 cell OSER containing lamellae and sinusoidal ER. F) Cos-7 cell branching SER interspersed with mitochondria (M). Inset features magnified image of the dense SER tubules. E and F images were generously provided by Drs. Maura Francolini and Nica Borgese. C and F from Snapp EL et al. J Cell Biol 2003; 163:257-269.[1]

to SER. Therefore, at steady-state, Sec61α is not equally distributed throughout the hepatocyte ER. Other translocon components (Ribophorins I and II, OST48, TRAP and Sec63) are also enriched in RER, but several translocon components do not differ in levels in the two domains (Sec61β, Dad1, Signal Peptidase, SRP receptor and TRAM). Surprisingly, some translocon components are present at high levels in SER (Ribophorin I and II), suggesting that these proteins may have

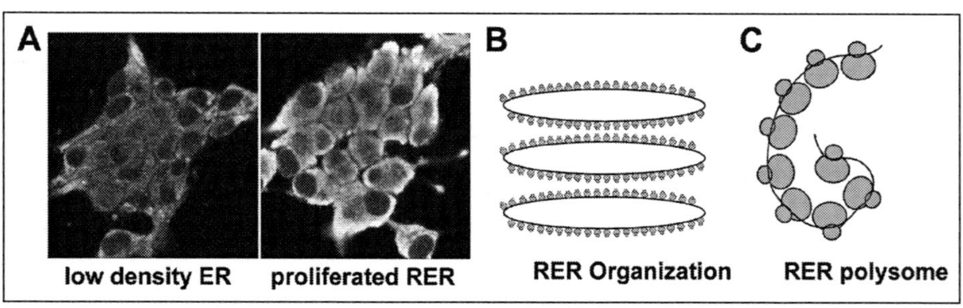

Figure 2. Rough ER at different resolutions. A) AR24J pancreatic acinar cells in a resting state (left panel) and stimulated with DMSO to proliferate RER (right panel). Cells were stained with anti-Sec61β and imaged by confocal microscopy. In addition to denser ER, the stimulated cells are also larger. B) Illustration of ribosome organization on cross-section of RER cisternae. In professional secretory cells, the RER cisternae are frequently closely apposed. C) Characteristic "G" shaped polysome organization that appears on RER cisternae.

Table 1. Translocon component distribution in rough and smooth ER

ER Protein	% Total RER Peptides	% Relative to Sec61α	% Total SER Peptides	% Relative to Sec61α
Sec61α	0.37	100.0	0.05	100.
Sec61β	0.09	23.8	0.09	192.1
Ribophorin I	1.28	346.0	0.55	1209.8
Ribophorin 2	1.16	312.0	0.61	1355.2
Dad1 (defender against cell death 1)	0.08	21.8	0.06	135.3
TRAPα	0.18	47.2	0.07	161.4
TRAPδ	0.26	70.0	0.11	236.3
TRAPγ	0.05	13.3	0.01	14.7
Signal peptidase 18kD subunit	0.04	11.3	0.05	102.4
Signal peptidase 25kD subunit	0.10	26.0	0.08	166.9
Sec63	0.08	20.6	0.00	0.00
SRP receptor α	0.01	1.5	0.01	27.3
SRP receptor β	0.06	17.5	0.09	200.3
TRAM1	0.01	1.5	0.01	12.6
BiP	1.41	380.1	1.36	3011.5
Calnexin	0.55	147.7	0.51	1131.5

Data are derived from Gilchrist et al.[12]

translocon-independent functions. The low levels of translocon components do not exclude the possibility of RER formation from SER. Black et al localized translocon components to purified SER of steroid-producing adrenal cells.[13] Steroid secreting cells have copious amounts of dense SER and a negligible amount of RER. In vitro translation assays using these SER microsomes revealed microsomes could translocate and glycosylate secretory protein substrates. However, these activities occurred at a much lower efficiency than in RER microsomes from canine pancreas, presumably due to the substantially lower levels of Sec61α and TRAP within the SER.

To visualize the localization of RER membrane proteins in living cells, Rolls et al used fluorescent protein (FP)-tagged RER membrane proteins in several cell types in live *C. elegans*.[14] In the majority of cell types tested, TRAM and SP12 (a signal peptidase subunit) colocalized with the general ER markers cytochrome p450 and phosphatidylinositol synthase. In neurons, which minimal SER, the RER markers (RAMP4, TRAPβ, TRAM and TRAPγ) were concentrated primarily in the cell body and depleted from neurites while the general ER markers localized to neurites as well. Their findings correlate with transmission EM micrographs confirming the concentration of ribosomes in neuronal cell bodies.

Another problem in RER formation is to determine how much of the Sec61 in cells is available to form new RER. Is Sec61 predominantly bound to ribosomes at steady state in cells? Two approaches have been employed to investigate this problem. Using biochemical fractionation techniques, MDCK cells were lysed in 500 mM potassium acetate and deoxy-BigCHAP and then fractionated on a sucrose gradient.[15] All detectable Sec61β was observed in the ribosome fractions, while the control ER membrane protein Calnexin was identified in the higher, ribosome-free fractions. Several other groups have reported similar observations.[16-18] In a second novel assay, Snapp et al [15] used immunofluorescence microscopy to

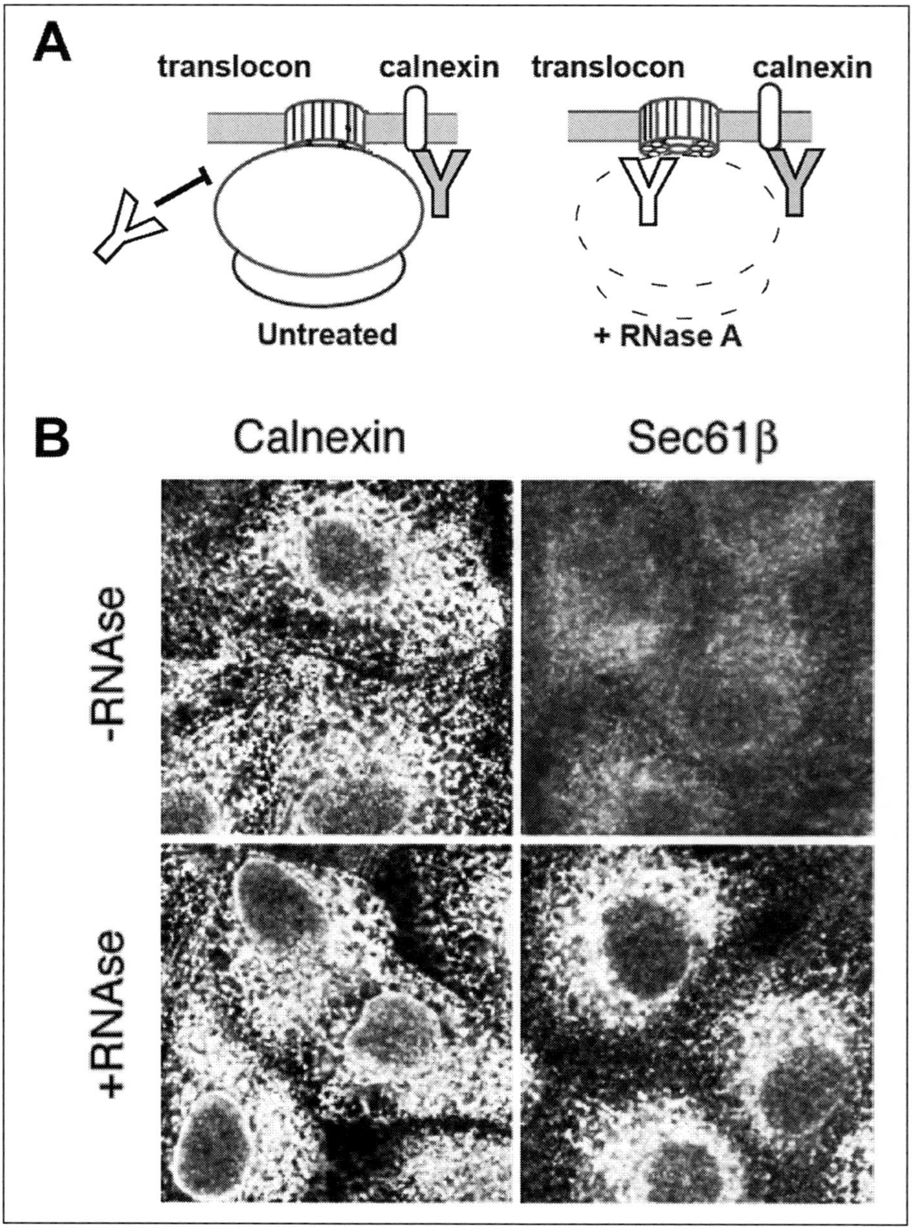

Figure 3. Translocon epitope accessibility assay. A) Illustration of assay. Cells are fixed and permeabilized. Antibodies can readily access cytoplasmic epitopes for nontranslocon proteins, such as the chaperone Calnexin. However, a ribosome-covered epitope is sterically obscured from interacting with an antibody. Treatment of the fixed permeabilized cells with RNaseA for 1 h digests much of the ribosome and now renders the translocon epitope accessible to the antibody. B) Immunofluorescence of cells untreated and treated with RNaseA. Calnexin staining is unaffected by RNase treatment. In contrast, staining of Sec61β is weak in untreated cells, but dramatically increases in intensity in RNase-treated cells. Image 3B from Snapp et al supplemental data. J Cell Biol 2004; 164:997-1007.[15]

directly visualize Sec61β association with ribosomes in intact cells. Using an antibody against the cytoplasmic domain of Sec61β, epitope accessibility was investigated in fixed cells. The Sec61β epitope labeled poorly in untreated cells (Fig. 3), whereas the cytoplasmic epitope on the ER membrane chaperone Calnexin labeled robustly. When ribosomes were partially digested with RNaseA treatment, Calnexin labeling remained unaffected, but the Sec61β labeling efficiency and intensity dramatically increased. The antibody epitope accessibility assay has considerable potential for investigating other ribosome-bound translocon proteins and their interactions with ribosomes under various conditions in cells. Together, the biochemical and fluorescence data suggest a low probability of ribosome binding to SER and indicate the majority of Sec61 is ribosome-bound at steady state.

Fractionation and fluorescence microscopy studies have shown that several core translocon proteins are substantially enriched in RER and ribosome-bound at steady state. The data collectively suggest the probability of forming new RER within existing SER is low due to the scarcity of RER-forming building blocks. In addition, SER biogenesis appears to be more complex than simply the absence of membrane bound ribosomes. Lipid synthesis, overexpression of integral membrane resident ER proteins and ER tubule-inducing proteins (i.e., reticulons) can modulate SER formation.[1,8,19-21] How RER and SER subdomains form and are maintained are active areas of study.

Translocon and Ribosome Mobility in Living Cells

Defining the steady state distribution of proteins does not specifically address the kinetics of how the steady state is achieved or maintained. To determine whether ribosomes and ribosome-bound Sec61 translocons readily sample the entire ER including SER domains, alternative imaging approaches were required. Standard fluorescence microscopy (i.e., fluorescence immunolocalization of proteins) lacks sufficient spatial resolution to distinguish individual ribosome-bound translocons in cells. This is due to a property of light, termed the Abbe' diffraction limit, which requires that similarly stained proteins must be ≥200 nm apart to be distinguishable as separate molecules. Ribosomes and the translocon are 25 nm and 10 nm in diameter respectively and ribosomes are frequently found immediately adjacent to one another, in RER at high density (Fig. 2). Transmission electron microscopy (TEM) easily resolves individual ribosomes on RER (Fig. 1), but TEM and immunofluorescence only provide static images of fixed cells. Fortunately, the cloning of green fluorescent protein (GFP)[22,23] and the development of photobleaching methods[24,25] have provided new tools for directly measuring protein dynamics in living cells.

Cells contain abundant ribosomes and according to one estimate, between 33-45% of total ribosomes are ER membrane-bound.[26] It has been unclear whether ribosome availability plays a role in RER biogenesis. Can cytoplasmic ribosomes easily encounter Sec61 in the ER? To address this question, Rolls et al expressed a fluorescent protein-tagged ribosome component, CFP-L23A to measure its mobility using fluorescence recovery after photobleaching (Fig. 4) (FRAP).[27,28] In FRAP, the FP of an FP-tagged protein is photobleached in a region of interest (ROI) using intense laser power and the rate of fluorescence recovery due to diffusion of unbleached molecules into the ROI is monitored with low laser power. The data obtained can then be analyzed for both the mobile fraction (the percent of unbleached proteins which contribute to fluorescence recovery into the ROI during the time of the experiment) and the diffusion coefficient (D) (a physical property of the molecule that depends on the size of the molecule and the viscosity of the environment).[27] Rolls et al performed FRAP on fluorescently labeled ribosomes in cells of live *C. elegans* and observed less than 10% fluorescence recovery into the bleach ROI over 90s, consistent with exceptionally low mobility. This FRAP experiment has several important implications for secretory protein synthesis.

Ribosomes appear to be restricted in their ability to sample the cytoplasm and encounter the ER. In practical terms, a translating ribosome must find an available translocon in close proximity to where translation is initiated. Proteins are translated at a rate of ~8 amino acids/second[29] and secretory proteins must interact with a translocon well before translation is completed (~1 minute for a typical 500 amino acid secretory protein). Results of a novel kinase accessibility assay

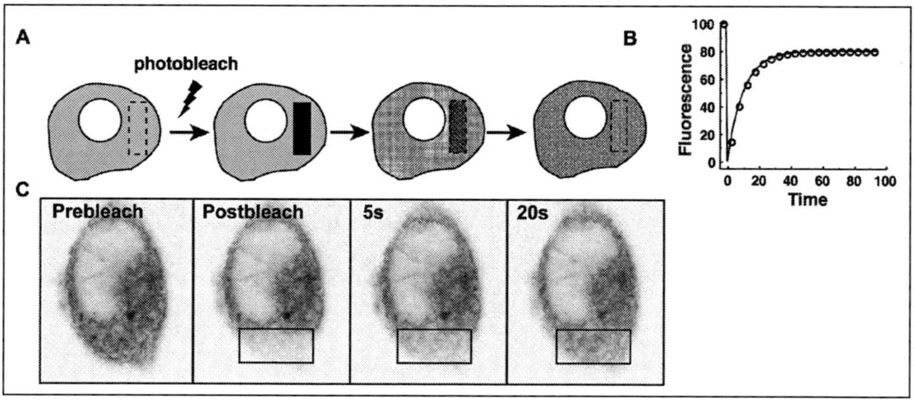

Figure 4. Fluorescence recovery after photobleaching (FRAP). A) Illustration of FRAP assay. An ROI in a fluorescently labeled cell is selectively photobleached and if unbleached fluorescent proteins are mobile, they will diffuse into the ROI and the whole cell will equilibrate to a dimmer overall fluorescence intensity. B) Plot of typical recovery of fluorescence intensity into photobleached ROI. The height of the asymptote of recovery indicates what fraction of fluorescent molecules in the cell can contribute to the fluorescence recovery. The shape of the recovery curve can be used to extract the diffusion coefficient of the fluorescent molecules. C) FRAP series of a living cell expressing an ER-localized GFP-tagged protein. Substantial recovery of fluorescence is apparent at 20s postbleach. The images have been inverted for ease of visualization.

by Goder et al indicated targeting of nascent chains to a translocon occurs within a few seconds.[29] Restricted ribosome mobility is not entirely consistent with this observation and suggests regulatory translational or targeting mechanisms must be present in cells. Lakkaraju et al recently reported evidence for translational slowing by Signal Recognition Particle and the formation of arrested polysomes ("ribosome stacking") in the cytoplasm and at translocons.[30] An alternative and complementary mechanism has been described by Potter and Nicchitta in which the majority of secretory protein synthesis initiates on translocons already bound to the 60S subunit of ribosomes.[18] After the initial assembly of a ribosome-bound translocon, only the mRNA and 40S subunit bind and release from the complex, which would circumvent the need for translating ribosomes to find an available translocon in a limited amount of time. Together, live cell and biochemical approaches highlight the importance of process efficiency, including ER targeting of translating ribosomes in the context of the crowded cellular environment.

Once RER and SER subdomains form, how stable are they? Rolls et al investigated whether RER membrane proteins were restricted in their mobility within cells. Using FRAP, Rolls et al determined the RER markers RAMP4 and SP12 are mobile,[14] indicating translocon components are not necessarily immobilized to or within the RER subdomain. A few caveats temper the interpretation of these FRAP studies. In particular, the study did not directly demonstrate functional incorporation of the proteins into translocons and did not quantitate the protein D values to determine the relative mobility of the translocon proteins.

Nikonov et al built on the observations of Rolls study by directly quantifying translocon mobility in living cells.[7] To perform these experiments, Nikonov et al generated the first functionally incorporated GFP-tagged translocon component, Dad1-GFP. This accomplishment is significant for two reasons. First, GFP is relatively large[31] (5 nm) and could sterically inhibit interactions with other protein partners.[32] For example, Snapp et al determined cytoplasmically oriented GFP-tagged Sec61 components failed to incorporate into translocons.[15] Second, competition with pre-existing untagged proteins and the frequently high levels of protein overexpression observed with transient transfection can adversely affect stoichiometric incorporation of GFP-tagged proteins into protein complexes. To circumvent these issues, Nikonov et al exploited a cell line expressing only a temperature sensitive mutant of Dad1, which rapidly degrades at the nonpermissive temperature.[7,33] Dad1, a 10 kD two-spanning transmembrane protein is essential for OST complex function and cell survival. At the nonpermissive temperature, stable GFP-Dad1 was functionally incorporated at levels comparable to native Dad1 into the OST complex and not observed independent of the OST complex. Nikonov et al exploited the cell line for FRAP analysis to determine the *D* of GFP-Dad1 and by extension, the translocon, which tightly associates with the OST complex.[34]

In actively translating cells, GFP-Dad1 diffused almost seven times slower (0.05 $\mu m^2/s$) than a typical ER membrane protein (0.3-0.7 $\mu m^2/s$).[1,5,27] To test the hypothesis that GFP-Dad1 exhibited low mobility due to translocon association with ribosomes assembled into polysomes (Fig. 2C), Nikonov et al perturbed the system with puromycin, which terminates nascent chain synthesis, thus resulting in polysome disassembly. GFP-Dad1 mobility increased three-fold ($D = 0.13\ \mu m^2/s$) after puromycin treatment, indicating a much higher mobility for single ribosome-translocon complexes relative to polysomes. The mobility of the inert ER membrane protein reporter remained unaffected. Conceptually, a three-fold difference in *D* indicates the slower diffusing molecule can only sample one ninth of the area of the more rapidly diffusing molecule in the same amount of time. By way of comparison, a polysome-translocon assembly can sample about one eightieth of the area of a typical ER membrane protein in the same time frame. Therefore, translation and polysome formation would serve to restrict translocons in their ability to disperse throughout the ER. We anticipate that levels of translation and the average size of transcripts or polysomes will significantly impact RER size, distribution and shape. However, this model does not incorporate barriers to the diffusion of large complexes.

In a follow-up study, Nikonov et al used the GFP-Dad1 cell line to assess microtubule involvement in the low mobility of the translocon.[7] Animal ER associates with microtubules and CLIMP-63, an RER-specific transmembrane protein, links RER membranes with microtubules through its cytoplasmic domain.[35] Membrane-bound microtubules would present significant barriers to polysome-associated translocons. FRAP was performed on cells under various microtubule altering conditions. The D of the translocon nearly doubled when cells were treated with the microtubule depolymerizing reagent nocodazole. In combination with puromycin treatment, D was not further increased with the addition of nocodazole. Conversely, treatment with the actin microfilament depolymerizer latrunculin B had no effect on GFP-Dad1 mobility. The lack of an effect of nocodazole in puromycin treated cells is consistent with the phenomenon of "sieving." Small molecules readily diffuse through spaces between large barriers while immense molecular complexes are obstructed. The interplay between secretory protein translation and physical barriers, including microtubules and possibly other oligomeric ER membrane proteins, such as reticulons,[36] is likely a major component of RER and SER subdomain formation and maintenance.

Translocon Composition and Dynamics

Eukaryotic translocons resemble transcription factor complexes in terms of the sheer number of proteins with varied functions that assemble into cellular machines. The mammalian Sec61 translocon is comprised of multiple (probably three to four) heterotrimeric Sec61 core complexes[15,37,38] composed of the conserved Sec61α, Sec61β and Sec61γ subunits.[16,34,38-40] The translocon core is complemented by several additional components essential to translocon function including the polytopic translocating chain-associating membrane protein (TRAM), the tetrameric translocon-associated protein complex (TRAPαβγδ), oligosaccharyl transferase (OST) complex (Ribophorin I and II, OST48, Dad1 and STTA/B), the Sec62 and Sec63 heterodimer, oligomeric signal peptidase (SP), RAMP4/Serp1 and the heterodimeric signal recognition particle receptor (SR).[6,41-45] Many secretory proteins require either TRAM or TRAP for efficient translocation into the ER lumen.[46-48] OST ensures correct N-glycosylation of nascent chains[49] and SP cleaves signal sequences from nascent chains.[50] RAMP4/Serp1 plays a role in the biogenesis and degradation of integral membrane proteins in conjunction with other translocon components.[51] Sec62 and Sec63 are implicated in posttranslational translocation of substrates in yeast.[52,53] However, their roles in metazoans remain unclear. As with transcription factors, it is likely a variety of translocons exist comprised of distinct components either by way of assembly through different regulatory processes or for performing unique functions. In yeast, two different translocons for co and posttranslational translocation have been described.[53-56] More recently, Sec61p translocon components have been implicated in ER associated degradation (ERAD).[57] It is likely mammalian translocon functions and heterogeneities are at least as complex as yeast translocons.

How are the twenty different translocon proteins organized in cells and how dynamic is their organization? Structural studies of purified translocons (see chapter by Wagner et al) have achieved atomic resolution for the bacterial equivalents of Sec61 and low nm resolution for mammalian tranlocons.[38] However, crystallographic and cryo-EM techniques, while powerful, are as yet limited in the types of questions that can be addressed regarding translocon structure. Crystallographic data typically describe static protein structures and previous crystallography studies have been limited to the bacterial translocon. Cryo-EM approaches require detergent extraction of translocons from their native membrane environment, potentially disrupting translocon interactions with several proteins relevant to translocation (i.e., TRAM and OST). In addition, cryo-EM methods average large populations of samples and could inadvertently obscure physiologically relevant heterogeneous assemblies. Currently, neither cryo-EM nor crystallographic approaches have been successful in addressing several fundamental questions regarding translocon organization in cells such as: How dynamic is the translocon? Are translocon components constantly in contact during all states of translocation? Are translocon component interactions relatively stable or more transient? How are translocon components spatially arranged during and after translocation? Is there more than one type of translocon and if so, are the different varieties heterogeneously distributed throughout the cell?

To investigate translocon component organization, Snapp et al exploited fluorescence resonance energy transfer (FRET) to assess relative proximities of native translocon components during the translocation cycle. FRET requires donor and acceptor fluorescent molecules, which exhibit a substantial overlap in the fluorescence emission of the donor and excitation spectra of the acceptor. Whereas fluorescence microscopy colocalization methods are typically restricted by diffraction to a maximal resolution of 200 nm, FRET can reveal whether colocalized molecules are within molecular proximities of each other. The principle of FRET is that when a donor fluorescent molecule is in close proximity (<10 nm) to an acceptor fluorophore, energy is transferred from the donor to the acceptor with increasing efficiency as the distance of the donor to the acceptor decreases. The consequences of FRET include an increase in the acceptor fluorophore signal when exciting the donor fluorophore and a decrease in the donor fluorophore emission. Of the several methods that can be employed to measure FRET; one of the most direct methods is acceptor photobleaching. Fluorescently labeled molecules are imaged in both the donor and acceptor channels. If the donor and acceptor are within FRET distances of each other, photobleaching the acceptor will de-quench the fluorescence signal from the donor (Fig. 5).

Initial attempts to label core translocon components with GFP were confounded by a failure of GFP-tagged translocon components to be incorporated into translocons.[15] As an alternative approach, anti-translocon component primary antibodies were labeled with donor (cy3) and acceptor (cy5) fluorescent dyes and fixed cells were stained with the antibodies. Despite the large size of IgG antibodies, (15 nm) the vast number of translocons in cells permitted detection of changes within protein proximities of 0.5 nm.

Using acceptor photobleaching antibody FRET, Snapp et al performed the first direct quantitation of the spatial relationship between several endogenously expressed translocon components in situ. The first part of the analyses focused on the translocon at steady state and revealed that the Sec61 heterotrimer assembles into an oligomer of two or more copies in cells. Several lines of evidence have suggested the translocon is composed of multiple Sec61 heterotrimers. Detection of FRET between two Sec61β proteins, as well as between two Sec61α proteins, confirmed this hypothesis.

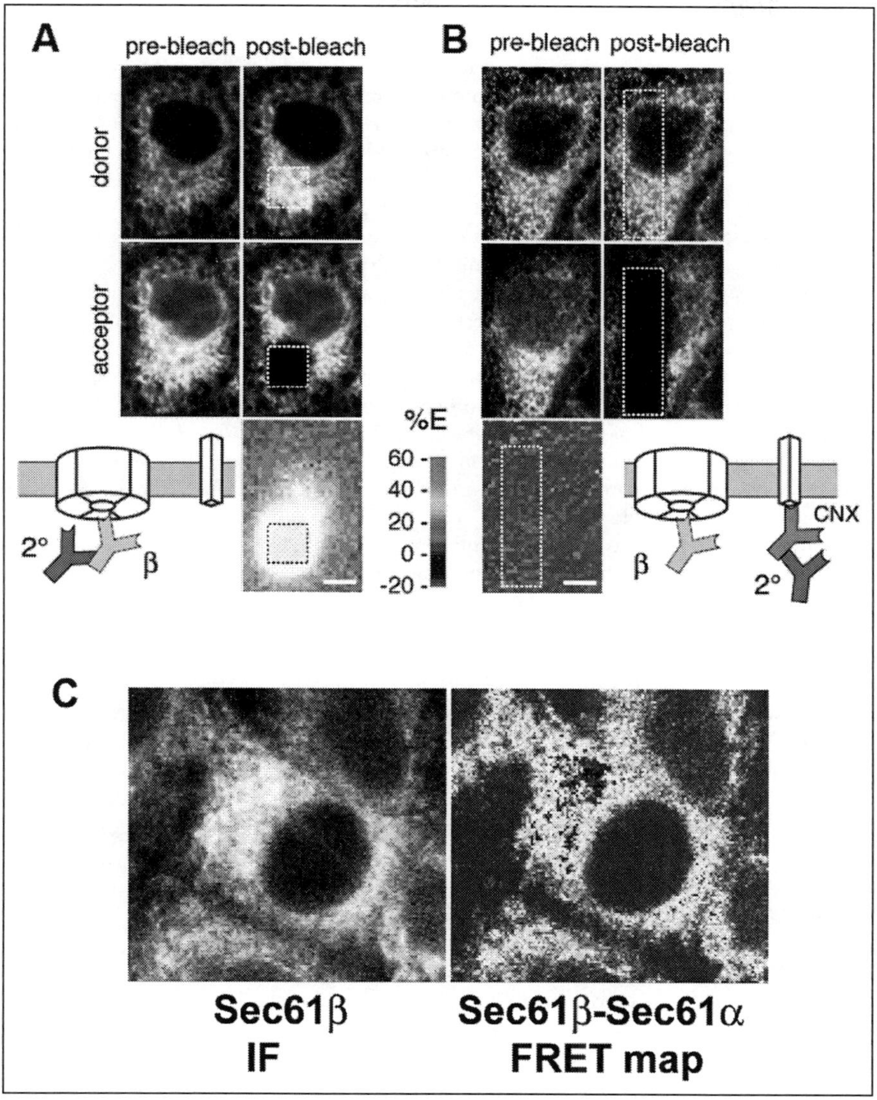

Figure 5. Fluorescence resonance energy transfer (FRET). A,B) Positive (A) and negative (B) controls for acceptor photobleaching FRET. Cells were fixed and labeled with an acceptor dye-labeled antibody against Sec61β and a secondary donor dye labeled antibody to either the Sec61β antibody or an unlabeled primary antibody against Calnexin. The acceptor antibody was photobleached in both samples (middle left panels). In A, where donor and acceptor are bound to each other, the donor intensity dramatically increased in the bleach region, while no change in donor intensity is apparent in B where donor and acceptor are significantly separated. The % Energy transfer efficiency maps quantitate FRET in the photobleached regions. A and B adapted from Snapp et al. J Cell Biol 2004; 164:997-1007.[15] C) Comparison of Sec61β immunofluorescence staining of a fixed MDCK cell (left panel) relative to the Sec61β donor to Sec61α acceptor FRET efficiency map of the whole cell after photobleaching the acceptor. Note how regions of FRET efficiency vary independently of the intensity of Sec61β staining. The variable FRET efficiency suggests that translocon organization and activity are spatially heterogeneous in cells.

Determining the organization of other mammalian translocon components has proven difficult by co-immunoprecipitation or similar purification methods.[16,47,58] Blue native PAGE and immunoblotting analysis of translocation components revealed SR and the OST complex could be found in multiple oligomeric states.[59] However, Sec61 was not detected with TRAP or OST components as part of the same Ribosome Associated Membrane Protein (RAMP) complex. In contrast, FRET in cells revealed that other proteins known to be important in the translocation process (TRAM, TRAPα and Signal Recognition Particle Receptor a) are also associated with the Sec61 translocon core at steady state. These data are consistent with more recent studies. A cryo-EM study using native ER membranes complete with translocon-docked ribosomes, suggested TRAP forms complexes with four Sec61 heterotrimers within the Sec61 translocon channel.[38] Also, RAMP analyses by Shibatani et al detected the association of Sec61, TRAP, signal peptidase and OST.[60]

FRET analysis also can detect subnanometer changes in the distance between translocon components. Next, Snapp et al asked whether translocon organization changes significantly between the presence and absence of nascent chain substrates. Cells were treated with one of two translational inhibitors (puromycin and pactamycin) and FRET was measured for several translocon components. The absence of gross changes in FRET values indicated that the core translocon remains intact throughout the translocation cycle. However, statistically significant changes in FRET were detected between translocon components, consistent with small allosteric changes in translocon organization upon engagement of substrate. An implication of this finding is that active and inactive translocons have FRET signatures and such signatures could be useful in detecting ER subdomains of translocon activity (Fig. 5C). Importantly, the data suggest that several translocons contain most if not all the translocon components necessary for the translocation of most secretory proteins, regardless of translocation activity. Thus, translocons would not need to constantly and repeatedly recruit and form interactions with the many translocon components. Furthermore, the necessity of RNaseA treatment for antibodies to be able to access translocon component epitopes revealed translocons are ribosome-bound in cells, even in the absence of nascent substrates (Fig. 3). From this, a model emerges from the FRET data in which fully assembled translocons remain intact and ribosome-bound throughout the translocation cycle. The persistence of preassembled and complete translocons between successive rounds of transport may facilitate highly efficient translocation in vivo. Preassembled translocons could help relieve the temporal constraints imposed by ongoing translation and the crowded cellular environment.

New Approaches for Translocon Studies in Cells

Several major questions remain unanswered in translocon physiology. Are translocons homogeneous in composition or are there multiple Sec61 translocons with distinct functions? Is translocon organization or composition different in distinct cellular subdomains or tissues? Does translocon organization or composition vary at points during the translocation cycle? Studying these questions in cells will require many of the methods discussed in this chapter, as well as new technologies described below.

Marking of Proteins in Cells

Since the cloning of GFP,[22] the number of new fluorescent proteins (FPs) has increased exponentially. Brighter and spectrally diverse proteins have become available to investigators.[61,62] Three relatively new classes of fluorescent proteins signify a new era of reagents that will prove to be highly beneficial to the advancement of translocon research. FlAsH, Photoactivatable (PA) and timer fluorescent proteins all permit optical marking of different populations of the same protein within the same individual cell or cellular compartment. Using these new tools, investigators can now perform optical pulse-chase experiments in single cells with up to sub-second temporal resolution in concert with exceptional spatial resolution. For each protein, the FP is fused to a protein of interest, as one would for GFP. FlAsH and ReAsH are small (17 amino acid) peptides that bind to exogenously added membrane permeant dyes.[63] The FlAsH system is a valuable alternative to the relatively large GFP and its potential for steric interference with translocon assembly or ribosome binding. In addition, ReAsH can be exploited to generate deposits of electron dense material in fixed samples to permit correlative electron microscopy, in conjunction with the fluorescence image of the fixed sample. The major advantage of FlAsH and ReAsH is that the green, red and now blue dyes can be added at different times to label a discrete population of proteins. Thus, one population can be labeled with one color and then all proteins synthesized in the next hour can be labeled with a different color.[63] Such an approach will permit the study of the fates of the translocon proteins and their RER distribution, specifically in terms of the changes in protein localization and turnover with time.

In contrast, PA proteins do not require addition of dyes but are instead induced to instantly undergo a color change with a pulse of light. PA proteins are derived from GFP and related fluorescent proteins and change from shorter to longer excitation and emission wavelengths upon light pulses. Investigators have several different types of PA proteins from which to choose.[61,64] A major advantage of PA proteins is that discrete regions or organelles within cells can be selectively activated. An even less invasive labeling method is available using timer proteins.[65] This class of fluorescent proteins also derived from GFP family members, changes color as the protein matures, typically converting from green to red. Therefore, the ratio of immature to mature fluorescent protein will reflect a combination of the rate of synthesis and the rate of turnover of a protein. The mature color also indicates whether older proteins exhibit a distinct distribution from newly synthesized proteins. The main drawback of timer proteins is the color change occurs over hours, whereas PA proteins and FlAsH peptides can be marked instantly or within minutes, respectively. A new, nonfluorescent protein marking method, TimeSTAMP, has been recently described and holds great potential for temporally marking proteins within whole animals.[66] The marked protein can then be labeled by standard immunofluorescence. All of these approaches will require genetically engineering the translocon-fusion proteins and as with the Nikonov et al study, the fusion proteins must then be verified as functionally integrated into translocons. These optical marking reagents hold great promise for studying translocon physiology in cells and animals.

Nanoscopy: Super-Resolution Microscopy

It would be of great utility to follow the behavior of single proteins or single translocons in cells. While the FRET approach is exceptionally powerful, it is generally employed as an ensemble method within cells. As previously mentioned, standard light microscopy cannot resolve fluorescently labeled proteins situated closer than 200 nm apart, well above the size of translocon components and translocons. The size of translocons and translocon components will necessitate new imaging technologies to resolve the protein composition and dynamics of individual translocons in cells. An important emerging technology, "nanoscopy" or so-called superresolution light microscopy, now permits resolution of molecules down to a range of 10-50 nm.

Since the late 1990s, cell structures and molecules have been resolved with superresolution methods that employ new microscope designs, new fluorescent probes and computer algorithms.[67] The most relevant superresolution microscopies for translocons and translocon components are STED and PALM. The two microscopies use different approaches to confine the signal emitted by a fluorescently labeled point source, termed the Point Spread Function (PSF). A property of light termed diffraction affects the ability of an observer to visualize a point-like fluorescent bead. Instead of detecting a tiny point, diffraction distorts the fluorescence emitted by the bead and yields an image of an oblong sphere that surrounds the true point source. Resolution depends on minimizing the size of the PSF. Large PSFs have to be spaced further apart to be resolved as separate PSFs. Very small PSFs permit the observer to distinguish small objects that are very close together. While a regular light microscope can produce a ~250 × 250 × 800 nm PSF and a confocal microscope can improve the PSF to ~200 × 200 × 500 nm, PALM and STED can optimally produce a PSF of 10 × 10 × 10 nm. In Photo-activated Localization Microscopy (PALM),[68,69] also termed Stochastic Optical Reconstruction Microscopy (STORM),[70,71] the PSF is not physically modified with a super microscope. Instead, individual fluorophores are visualized over the entire image field, a few at a time. Visualizing single molecules is achieved by employing PA fluorophores that are initially turned off and are then randomly turned on a few at a time. The activated fluorophores are imaged until they photobleach. Then a few more fluorophores are activated successively until all of the fluorophores in the field have been photoactivated and photobleached to assemble the image. The other key component of PALM is to localize each individual fluorophore with high precision by performing a Gaussian fitting routine to pinpoint the center of the PSF for each fluorophore.[72,68] Currently, PALM remains best suited to study immobilized molecules. Additionally, when PALM is coupled with TEM, the distributions of individual molecules can be determined on or within an organelle. Such correlative microscopy methods should be exceptionally useful for studying the organization and composition of large molecular assemblies, such as translocons.

Stimulated Emission Depletion (STED) employs optical methods to achieve superresolution. Two laser pulses are used to produce a signal for each pixel of an image. The first nanosecond pulse is focused into a confocal-sized volume and the laser is tuned to the absorption wavelength of the fluorophore of interest. A second intense multiphoton donut-shaped laser pulse immediately follows the first pulse and is tuned to the fluorophore emission wavelength. The second pulse stimulates excited fluorophores to fall back to the ground state and not emit any photons in the depletion pulse. Only fluorophores in the outer part of the diffraction-limited spot are depleted, while the small volume of molecules in the donut center still fluoresce. The size of the donut directly regulates the PSF and can improve resolution down to 10-20 nm. More recently, Stefan Hell and colleagues have achieved video rate (28 frames per second or about 35 ms/frame) imaging of movement of 50 nm vesicles within a neuronal synapse.[73] Fortunately, because translocons diffuse so slowly, such high speeds may not be necessary for translocon studies.

An alternative to superresolution microscopy techniques is to count individual fluorophores in cells. Fluorescence Correlation Spectroscopy (FCS) utilizes a confocal or multiphoton microscope setup to define a femtoliter volume (~200 nm × 500 nm). Using this setup, an exceptionally sensitive detector then counts photons emitted from fluorophores as they pass through the volume. The rate of movement of the molecules through the volume and number of fluorophores counted can then be analyzed by a Fourier transformation to derive D values and concentrations of species that diffuse at significantly different rates. In practical terms, this means one could determine the percent of translocon-incorporated and free proteins in a subdomain of the cell. One could also use two different color tagged-proteins and perform cross-correlation analysis to directly measure the fraction of complexes that contain both proteins versus free proteins. FCS can be applied to address many of the same questions as FRAP and FRET, but incorporates exceptional temporal resolution with the ability to measure protein concentrations. The latter feature can also be coupled with cross-correlation analysis to measure protein on and off rates, thus permitting an investigator to perform biochemistry in single living cells.

Concluding Remarks

Live cell biophysical fluorescence microscopy methods are powerful new tools for translocon studies. Many of the technologies described in this chapter are currently accessible to researchers at their own institutions or through collaborations. When coupled with traditional biochemical methods, biophysical fluorescence imaging methods will shift research from a steady-state view of translocons to a dynamic view of translocons and translocon components. These approaches will produce new insights into how protein localization, concentration, diffusion and interactions regulate translocon function within living cells.

References

1. Snapp EL, Hegde RS, Francolini M et al. Formation of stacked ER cisternae by low affinity protein interactions. J Cell Biol 2003; 163:257-269.
2. Borgese N, Francolini M, Snapp E. Endoplasmic reticulum architecture: Structures in flux. Curr Opin Cell Biol 2006; 18:358-364.
3. Baumann O, Walz B. Endoplasmic reticulum of animal cells and its organization into structural and functional domains. Int Rev Cytol 2001; 205:149-214.

4. Deng Y, Bennink JR, Kang HC et al. Fluorescent conjugates of Brefeldin a selectively stain the endoplasmic reticulum and Golgi complex of living cells. J Histochem Cytochem 1995; 43:907-915.
5. Ellenberg J, Siggia ED, Moreira JE et al. Nuclear membrane dynamics and reassembly in living cells: Targeting of an inner nuclear membrane protein in interphase and mitosis. J Cell Biol 1997; 138:1193-1206.
6. Kreibich G, Ulrich BL, Sabatini DD. Proteins of rough microsomal membranes related to ribosome binding. I. Identification of Ribophorins I and II, membrane proteins characteristics of rough microsomes. J Cell Biol 1978; 77:464-487.
7. Nikonov AV, Snapp E, Lippincott-Schwartz J et al. Active translocon complexes labeled with GFP-Dad1 diffuse slowly as large polysome arrays in the endoplasmic reticulum. J Cell Biol 2002; 158:497-506.
8. Shibata Y, Voeltz GK, Rapoport TA. Rough sheets and smooth tubules. Cell 2006; 126:435-439.
9. Morrow MW, Brodsky JL. Yeast ribosomes bind to highly purified reconstituted Sec61p complex and to mammalian p180. Traffic 2001; 2:705-716.
10. Becker F, Block-Alper L, Nakamura G et al. Expression of the 180-kD ribosome receptor induces membrane proliferation and increased secretory activity in yeast. J Cell Biol 1999; 146:273-284.
11. Bai JZ, Leung E, Holloway H et al. Alternatively spliced forms of the p180 ribosome receptor differ in their ability to induce the proliferation of rough endoplasmic reticulum. Cell Biol Int 2008; 32:473-483.
12. Gilchrist A, Au CE, Hiding J et al. Quantitative proteomics analysis of the secretory pathway. Cell 2006; 127:1265-1281.
13. Black VH, Sanjay A, van Leyen K et al. Cholesterol and steroid synthesizing smooth endoplasmic reticulum of adrenocortical cells contains high levels of proteins associated with the translocation channel. Endocrinology 2005; 146:4234-4249.
14. Rolls MM, Hall DH, Victor M et al. Targeting of rough endoplasmic reticulum membrane proteins and ribosomes in invertebrate neurons. Mol Biol Cell 2002; 13:1778-1791.
15. Snapp EL, Reinhart GA, Bogert BA et al. The organization of engaged and quiescent translocons in the endoplasmic reticulum of mammalian cells. J Cell Biol 2004; 164:997-1007.
16. Gorlich D, Prehn S, Hartmann E et al. A mammalian homolog of Sec61p and SecYp is associated with ribosomes and nascent polypeptides during translocation. Cell 1992; 71:489-503.
17. Matlack KE, Walter P. The 70 carboxyl-terminal amino acids of nascent secretory proteins are protected from proteolysis by the ribosome and the protein translocation apparatus of the endoplasmic reticulum membrane. J Biol Chem 1995; 270:6170-6180.
18. Potter MD, Nicchitta CV. Endoplasmic reticulum-bound ribosomes reside in stable association with the translocon following termination of protein synthesis. J Biol Chem 2002; 277:23314-23320.
19. Voeltz GK, Prinz WA. Sheets, ribbons and tubules—how organelles get their shape. Nat Rev Mol Cell Biol 2007; 8:258-264.
20. Sprocati T, Ronchi P, Raimondi A et al. Dynamic and reversible restructuring of the ER induced by PDMP in cultured cells. J Cell Sci 2006; 119:3249-3260.
21. Szczesna-Skorupa E, Chen CD, Liu H et al. Gene expression changes associated with the endoplasmic reticulum stress response induced by microsomal cytochrome p450 overproduction. J Biol Chem 2004; 279:13953-13961.
22. Chalfie M, Tu Y, Euskirchen G et al. Green fluorescent protein as a marker for gene expression. Science 1994; 263:802-805.
23. Cubitt AB, Heim R, Adams SR et al. Understanding, improving and using green fluorescent proteins. Trends Biochem Sci 1995; 20:448-455.
24. Velez M, Axelrod D. Polarized fluorescence photobleaching recovery for measuring rotational diffusion in solutions and membranes. Biophys J 1988; 53:575-591.
25. Lippincott-Schwartz J, Smith CL. Insights into secretory and endocytic membrane traffic using green fluorescent protein chimeras. Curr Opin Neurobiol 1997; 7:631-639.
26. Rajasekaran AK, Langhans-Rajasekaran SA, Gould RM et al. A simple biochemical approach to quantitate rough endoplasmic reticulum. Am J Physiol 1995; 268:C308-316.
27. Snapp EL, Altan N, Lippincott-Schwartz J. Measuring protein mobility by photobleaching GFP chimeras in living cells. Curr Protoc Cell Biol 2003; Unit 21.1.
28. Lippincott-Schwartz J, Snapp E, Kenworthy A. Studying protein dynamics in living cells. Nat Rev Mol Cell Biol 2001; 2:444-456.
29. Goder V, Crottet P, Spiess M. In vivo kinetics of protein targeting to the endoplasmic reticulum determined by site-specific phosphorylation. EMBO J 2000; 19:6704-6712.
30. Lakkaraju AK, Mary C, Scherrer A et al. SRP keeps polypeptides translocation-competent by slowing translation to match limiting ER-targeting sites. Cell 2008; 133:440-451.
31. Yang F, Moss LG, Phillips GN Jr. The molecular structure of green fluorescent protein. Nat Biotechnol 1996; 14:1246-1251.

32. Andresen M, Schmitz-Salue R, Jakobs S. Short tetracysteine tags to beta-tubulin demonstrate the significance of small labels for live cell imaging. Mol Biol Cell 2004; 15:5616-5622.
33. Nishimoto T, Basilico C. Analysis of a method for selecting temperature-sensitive mutants of BHK cells. Somatic Cell Genet 1978; 4:323-340.
34. Menetret JF, Neuhof A, Morgan DG et al. The structure of ribosome-channel complexes engaged in protein translocation. Mol Cell 2000; 6:1219-1232.
35. Nikonov AV, Hauri HP, Lauring B et al. CLIMP-63-mediated binding of microtubules to the ER affects the lateral mobility of translocon complexes. J Cell Sci 2007; 120:2248-2258.
36. Shibata Y, Voss C, Rist JM et al. The reticulon and DP1/Yop1p proteins form immobile oligomers in the tubular endoplasmic reticulum. J Biol Chem 2008; 283:18892-18904.
37. Morgan DG, Menetret JF, Neuhof A et al. Structure of the mammalian ribosome-channel complex at 17Å resolution. J Mol Biol 2002; 324:871-886.
38. Menetret JF, Hegde RS, Heinrich SU et al. Architecture of the ribosome-channel complex derived from native membranes. J Mol Biol 2005; 348:445-457.
39. Hartmann E, Sommer T, Prehn S et al. Evolutionary conservation of components of the protein translocation complex. Nature 1994; 367:654-657.
40. Hanein D, Matlack KE, Jungnickel B et al. Oligomeric rings of the Sec61p complex induced by ligands required for protein translocation. Cell 1996; 87:721-732.
41. Hegde RS, Voigt S, Rapoport TA et al. TRAM regulates the exposure of nascent secretory proteins to the cytosol during translocation into the endoplasmic reticulum. Cell 1998; 92:621-631.
42. Hartmann E, Gorlich D, Kostka S et al. A tetrameric complex of membrane proteins in the endoplasmic reticulum. Eur J Biochem 1993; 214:375-381.
43. Kelleher DJ, Kreibich G, Gilmore R. Oligosaccharyltransferase activity is associated with a protein complex composed of Ribophorins I and II and a 48 kD protein. Cell 1992; 69:55-65.
44. Halic M, Gartmann M, Schlenker O et al. Signal recognition particle receptor exposes the ribosomal translocon binding site. Science 2006; 312:745-747.
45. Jiang Y, Cheng Z, Mandon EC et al. An interaction between the SRP receptor and the translocon is critical during cotranslational protein translocation. J Cell Biol 2008; 180:1149-1161.
46. Gorlich D, Hartmann E, Prehn S et al. A protein of the endoplasmic reticulum involved early in polypeptide translocation. Nature 1992; 357:47-52.
47. Fons RD, Bogert BA, Hegde RS. Substrate-specific function of the translocon-associated protein complex during translocation across the ER membrane. J Cell Biol 2003; 160:529-539.
48. Voigt S, Jungnickel B, Hartmann E et al. Signal sequence-dependent function of the TRAM protein during early phases of protein transport across the endoplasmic reticulum membrane. J Cell Biol 1996; 134:25-35.
49. Silberstein S, Gilmore R. Biochemistry, molecular biology and genetics of the oligosaccharyltransferase. FASEB J 1996; 10:849-858.
50. Blobel G, Dobberstein B. Transfer of proteins across membranes. I. Presence of proteolytically processed and unprocessed nascent immunoglobulin light chains on membrane-bound ribosomes of murine myeloma. J Cell Biol 1975; 67:835-851.
51. Yamaguchi A, Hori O, Stern DM et al. Stress-associated endoplasmic reticulum protein 1 (SERP1)/Ribosome-Associated Membrane Protein 4 (RAMP4) stabilizes membrane proteins during stress and facilitates subsequent glycosylation. J Cell Biol 1999; 147:1195-1204.
52. Deshaies RJ, Schekman R. Sec62 encodes a putative membrane protein required for protein translocation into the yeast endoplasmic reticulum. J Cell Biol 1989; 109:2653-2664.
53. Rothblatt JA, Deshaies RJ, Sanders SL et al. Multiple genes are required for proper insertion of secretory proteins into the endoplasmic reticulum in yeast. J Cell Biol 1989; 109:2641-2652.
54. Deshaies RJ, Schekman R. A yeast mutant defective at an early stage in import of secretory protein precursors into the endoplasmic reticulum. J Cell Biol 1987; 105:633-645.
55. Schnell DJ, Hebert DN. Protein translocons: Multifunctional mediators of protein translocation across membranes. Cell 2003; 112:491-505.
56. Green N, Fang H, Walter P. Mutants in three novel complementation groups inhibit membrane protein insertion into and soluble protein translocation across the endoplasmic reticulum membrane of Saccharomyces cerevisiae. J Cell Biol 1992; 116:597-604.
57. Scott DC, Schekman R. Role of Sec61p in the ER-associated degradation of short-lived transmembrane proteins. J Cell Biol 2008; 181:1095-1105.
58. Gorlich D, Rapoport TA. Protein translocation into proteoliposomes reconstituted from purified components of the endoplasmic reticulum membrane. Cell 1993; 75:615-630.
59. Wang L, Dobberstein B. Oligomeric complexes involved in translocation of proteins across the membrane of the endoplasmic reticulum. FEBS Lett 1999; 457:316-322.

60. Shibatani T, David LL, McCormack AL et al. Proteomic analysis of mammalian oligosaccharyltransferase reveals multiple subcomplexes that contain Sec61, TRAP and two potential new subunits. Biochemistry 2005; 44:5982-5992.
61. Shaner NC, Patterson GH, Davidson MW. Advances in fluorescent protein technology. J Cell Sci 2007; 120:4247-4260.
62. Shaner NC, Steinbach PA, Tsien RY. A guide to choosing fluorescent proteins. Nat Methods 2005; 2:905-909.
63. Gaietta G, Deerinck TJ, Adams SR et al. Multicolor and electron microscopic imaging of connexin trafficking. Science 2002; 296:503-507.
64. Lukyanov KA, Chudakov DM, Lukyanov S et al. Innovation: Photoactivatable fluorescent proteins. Nat Rev Mol Cell Biol 2005; 6:885-891.
65. Terskikh A, Fradkov A, Ermakova G et al. "Fluorescent timer": Protein that changes color with time. Science 2000; 290:1585-1588.
66. Lin MZ, Glenn JS, Tsien RY. A drug-controllable tag for visualizing newly synthesized proteins in cells and whole animals. Proc Natl Acad Sci USA 2008; 105:7744-7749.
67. Hell SW. Far-field optical nanoscopy. Science 2007; 316:1153-1158.
68. Betzig E, Patterson GH, Sougrat R et al. Imaging intracellular fluorescent proteins at nanometer resolution. Science 2006; 313:1642-1645.
69. Hess ST, Girirajan TP, Mason MD. Ultra-high resolution imaging by fluorescence photoactivation localization microscopy. Biophys J 2006; 91:4258-4272.
70. Huang B, Wang W, Bates M et al. Three-dimensional super-resolution imaging by stochastic optical reconstruction microscopy. Science 2008; 319:810-813.
71. Rust MJ, Bates M, Zhuang X. Sub-diffraction-limit imaging by Stochastic Optical Reconstruction Microscopy (STORM). Nat Methods 2006; 3:793-795.
72. Park H, Toprak E, Selvin PR. Single-molecule fluorescence to study molecular motors. Q Rev Biophys 2007; 40:87-111.
73. Westphal V, Rizzoli SO, Lauterbach MA et al. Video-rate far-field optical nanoscopy dissects synaptic vesicle movement. Science 2008; 320:246-249.

CHAPTER 4

Structure of the Sec61-Complex

Richard Wagner,* Frank Erdmann, Anke Harsman and Alf Honigmann

Abstract

A conserved protein translocation channel is formed by the hetero-trimeric Sec61 complex in eucaryotes and by SecY complex in archaea and eubacteria. Experimental observations including biochemical, biophysical and electron microscopy studies indicate that oligomers of the hetero-trimeric Sec61/SecY complexes form the channel through the membrane. Nevertheless, the determined crystallographic high resolution structure of an archaeal SecY complex indicates that the pore may be located at the center of a single hetero-trimeric complex. Based on this structure molecular models were developed to envisage how secretory proteins are transported across the membrane and how lateral insertion of transmembrane segments into the lipid bilayer may occur.

Introduction

Most of the proteins of a cell are synthesized in the cytoplasm and about half of them are inserted into or translocated across membranes.[1] These processes are mediated by hydrophilic membrane pores, identified as translocons. First evidence for a protein conducting channel in pancreatic rough microsomal (RM) vesicles came from the electrophysiological experiments by Simon and Blobel, 1991.[2] Protein translocons consist of complex molecular machines possessing essential features required for protein targeting into and across biomembranes.[3] These features include intrinsic signal recognition sites facilitating docking of targeting signals in nascent polypeptide chains as well as the formation of selectively permeable protein-conducting membrane channels that mediate transport from one (cis) compartment to the destination (trans) compartment. Finally, translocons must provide the driving force for the unidirectional transport of polypeptides either by coupling to endogenous energy providing complexes or by association of molecular chaperones with the polypeptide in the trans compartment. Compared to other membrane transporters, translocons represent a class of remarkably flexible transport complexes. The channel pores of translocons can accommodate a multitude of distinct protein substrates (from dipeptides to >1MDa polypeptides) while maintaining the permeability barrier of the membrane. In addition to providing a hydrophilic channel for complete translocation of polypeptides, many translocons sense stop-transfer signals within integral membrane proteins and open laterally to allow the release of transmembrane segments into the membrane (see chapter by Cross and High). Therefore, the channels do not only supply a passive conduit but are actively participating in the translocation process. Studies on different protein translocation pores indicate that nascent or newly synthesized polypeptides are transported vectorially through a regulated protein-conducting channel in a mainly unfolded conformation with the aid of molecular chaperones.[3,4]

By maintaining the polypeptide in a mainly unfolded and thus rather unique conformation, a single translocon of defined topology can accommodate a huge collection of different substrates. From this, the translocation process can be modeled as the transport of poly-ions through signal-gated translocon channels.

*Corresponding Author: Richard Wagner—Biophysics, Department of Biology/Chemistry, University Osnabrück, 49076 Osnabrück, Germany. Email: wagner@uos.de

Protein Transport into the Endoplasmic Reticulum, edited by Richard Zimmermann.

Transport Modes of the Sec61/SecY Channel

Protein translocation across and insertion into the membrane of the endoplasmic reticulum membrane in eucaryotes or the plasma membrane in procaryotes is mediated by the Sec61/SecY complex. The predominant pathway in higher eucaryotes is the cotranslational translocation, during which the nascent unfolded polypeptide chain is synthesized by a membrane-bound ribosome and simultaneously translocated across the membrane. During posttranslational translocation, used extensively in yeast and bacteria, the ribosome is absent from the translocation complex and the mainly unfolded polypeptide is translocated after completion of translation. Additional proteins like BiP in eucaryotes or SecA in bacteria associate with the Sec complex exerting important additional functions like generation of driving force. Furthermore the Sec61 complex appears to play a major role in retro-translocation of misfolded proteins or bacterial toxins from the ER-lumen to the cytosol.[5,6] In this transport mode also folded or partially folded polypeptides seem to be transported across the Sec61 pore as well.[7,8]

Demands on a Protein Conducting Channel

In order to fulfill co and posttranslational translocation as well as insertion of proteins the translocon has to apply certain structural demands. The complex has to provide a water-filled channel, large and hydrophilic enough to transport polypeptide chains in the unfolded,[9] mainly unfolded[8] and partially folded state.[10] It must identify the signal sequences of approaching substrates and open in response to them. Besides the translocation of soluble secretory proteins, the Sec complex also manages membrane protein translocation and integration. This necessitates the ability to open laterally towards the lipid phase and to release hydrophobic domains of preproteins that will ultimately reside in the membrane. Furthermore it has to sort out those loops and domains of the protein determined to stay in the cytosol and prevent them from being translocated.

All of these steps have to be performed while the membrane barrier is kept intact, as the endoplasmatic reticulum plays a vital role as dynamic calcium reservoir and in the propagation of cytosolic calcium waves during intracellular signalling pathways.[11] In bacteria the translocon also has to maintain the electrochemical proton gradient across the plasma membrane. In addition the complex has to provide binding sites to other components of the translocon machinery, such as the ribosome or the molecular ratchet BiP.

Deciphering how all these demands on structure and function are accomplished at the molecular level has been a central goal to researchers during the past decade.[12]

History

Many efforts have been implied in unravelling the structure of the hetero-trimeric Sec-complex, seeking to gain a deeper understanding of the structure-function relationship. In eucaryotes the complex is composed of Sec61α, spanning the membrane ten times with the n- and c-termini exposed to the cytosol and two smaller subunits Sec61γ and Sec61β containing one transmembrane segment each.[13,14]

The first low resolution structures of membrane proteins associated with ribosomes in the rough endoplasmic reticulum came from freeze-fracture electron microscopy (EM) of rough microsomes prepared from rat liver.[15] Analog structures showing membrane particles with a diameter of about 100 Å could be visualized in ER membranes derived from the green alga *Micrasterias denticulata*.[16] Comparable dimensions were observed with negatively-stained, purified Sec complexes from mammals, yeast[17,18] and bacteria[18] in detergent. However, these pictures showed a central, stain filled cavity with a diameter of 20 Å giving the complex a donut-like structure large enough to contain 3-4 Sec61 hetero-trimers.

The first three-dimensional reconstruction (26 Å resolution) of the ribosome-Sec61 complex in detergent based on cryo-electron microscopy was published by Beckmann et al.[19] It showed the Sec61 oligomer bridged to the ribosome by a single connection. The reconstruction also revealed a toroidal structure, representing the translocase, whose inner diameter ranged from 15 to 35 Å. This pore structure aligns with the ribosomal exit tunnel supporting the idea that during cotranslational translocation a nascent chain can directly elongate from the ribosome into the translocation channel.[19] Subsequent reconstructions of the ribosome-channel complexes in absence and presence of a nascent polypeptide provided a similar picture, differing only in the number of connections coupling ribosome and Sec complex.[20] However, the central pore postulated till then could not be confirmed in the following reconstructions at 15.4 Å and 17 Å, but was likely to be a small indentation, a difference probably due to refined contour levels.[21,22] Concurrently Collinson et al published a higher resolution structure of the *E. coli* SecYEG from 2-D crystals containing the hetero-trimeric complex.[23] The 3-D map (8 Å resolution) calculated from cryo-EM data of the 2-D crystals revealed a double layer of membranes interacting through the large cytoplasmic loops of SecY. All of the 15 α-helices predicted for the *E. coli* hetero-trimer[24] could be allocated to densities probably representing a SecYEG monomer. In these crystals the monomers were assembled into clearly separated dimers of SecYEG that comprise a 22 Å deep cavity of 16 × 25 Å width at their interface which is closed towards the periplasmic side.[25] In 2004 a high-resolution crystal structure (3.2 Å) of the SecY complex from the archaean *Methanococcus jannaschii* was presented. It showed a single hetero-trimer and revealed features promoting the idea that the monomer may serve as functional translocation channel with the structure probably representing the closed state.[26]

X-Ray Structure

The X-ray structure of the protein conducting channel from *M. jannaschii* solved by van den Berg et al[26] reveals that the 10 trans-membrane helices of SecY group into two domains, TM1-TM5 and TM6-TM10 with the amino- and carboxy-termini exposed to the cytosol (Fig. 1). The helices TM1-TM5 and TM5-TM10 form opposing bundles resulting in a clamshell-like shape with SecE forming a clamp around the back of SecY (Fig. 1A). The two halves of the clamshell are separated by an incomplete ring of hydrophobic residues. In contrast to earlier observations with the *E. coli* SecYEG complex[25] the crystal structure from *M. jannaschii* SecYEβ shows that a cavity with a diameter of about 20 Å is formed within a monomer. Beginning at the cytoplasmic face of SecY it tapers to a close in the centre of the membrane. A second vestibule-like structure extends from the periplasmic side to the center of the complex resulting in an hourglass-like shape. It has been postulated that the channel-like structure resulting from the connection of these two vestibules

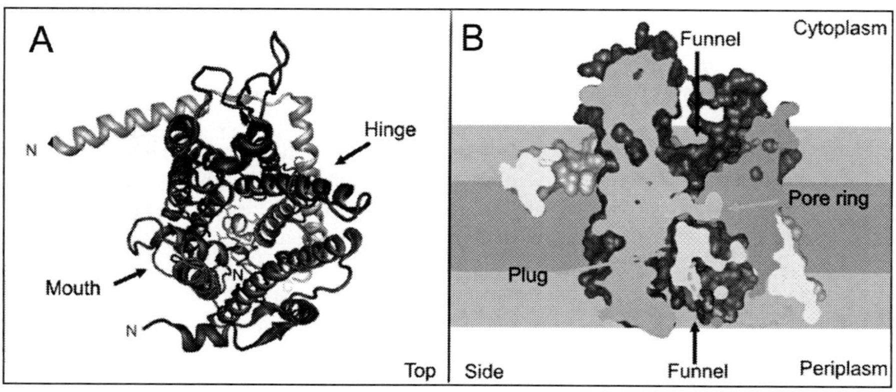

Figure 1. X-ray struture of SecYEβ from M. jannaschii. A) Ribbon representation of the SecY complex viewed from the cytoplasm. SecY is coloured in red (TM1-TM5) and blue (TM6-TM10), the plug (TM2a) is coloured orange. Green sticks show the hydrophobic pore ring residues. SecE and Secβ are coloured purple and yellow, respectively. B) Side view of the SecY complex (surface rendering) sliced through the middle with the membrane coloured in grey. Colour scheme as in a. Indicated are the cytoplasmic and exctracellular funnels as well as proposed plug movement (dashed orange line). Figures were adapted from Clemons et al, 2004.[27] A color version of this image is available at www.landesbioscience.com/curie.

might represent the path for the translocating polypeptide, suggesting that a single copy of the SecYEβ heterotrimer forms the translocation pore.[26]

However, the periplasmic funnel is disguised by the small helix TM2a, which extends from the extracellular side into the channel thus plugging periplasmic access, indicating that the structure corresponds to the closed channel. Opening of the channel could occur by a movement of the plug towards the external side of the complex. This suggestion was supported by crosslinking experiments which demonstrated that TM2a can be linked to the c-terminal end of SecE even though theses residues are 20 Å apart in the X-ray structure.[28,29] At its narrowest constriction the channel is lined by six hydrophobic residues (mainly isoleucines), forming a ring with 5-8 Å diameter (center-of-atom to center-of-atom distance). Considering the real occupied volumes of these residues, the according helices would need to move in a concerted way with large amplitudes to result in a stretch of the ring thus forming a pore that allows passage of nascent chains. The additional displacement of the "plug" helix TM2a would open the aqueous translocation conduit.

This would explain how in vitro translocation of polypeptides with bulky groups attached to lysine side chains[29] or proOmpA containing a disulfide-bonded loop of 13 amino acids[10,30] can be conducted by a single hetero-trimer. Besides the accessibility for differently sized substrates, the elasticity and hydrophobicity of the pore ring may result in a gasket-like seal around a translocating polypeptide thereby preventing the passage of ions.

In this model the front of the clamshell could account for the lateral release of hydrophobic transmembrane segments of nascent membrane proteins. As the X-ray structure shows, the interface formed by TM2b, TM3, TM7 and TM8 is the only site of the complex capable of opening laterally towards the lipid phase.[26]

The combined action of plug and pore ring may also be an explanation for how the channel is able to maintain a membrane barrier for small molecules in the active as well as in the nontranslating state. Electrophysiological measurements with plug-deletion and prl mutants provided supplementary evidence for plug and pore ring as major contributors to the permeation barrier. A double mutation with cysteines introduced into the plug and the side of the SecE subunit where the plug rests in the open conformation of the channel resulted in an open channel after disulfide bridge formation.[31] Additionally, studies using plug-deletion mutants recently gave further evidence on the importance of the plug for the stabilization of the closed channel and thereby the maintenance of the membrane barrier.[32]

Oligomerization

Although certain features of the *M. jannaschii* SecYEβ crystal structure suggest that translocation of polypeptides may be conducted by a single Sec hetero-trimer,[26] the majority of additional data indicate that the Sec complex is constituted by oligomers of the hetero-trimer in vivo. Early negative stain electron microscopy studies of native ER membranes and purified, liposome reconstituted Sec61 complexes revealed that ring-like structures with a diameter of up to 11 nm assemble in lipid bilayers. The number of ring-like structures in liposomes increased when channel partners were coreconstituted indicating that oligomerization may be facilitated by additional components such as Sec62/63 or the ribosome.[16,17] Similar structures were found with lipid and detergent reconstituted SecYE from *Bacillus subtilis.*[17,18] All ring-like structures had a central stain filled cavity which was thought to represent the protein conducting channel at that time. The number of hetero-trimers in the oligomeric rings was estimated to be 3-4.

The Sec61 complex bound to a ribosome with and without a nascent polypeptide chain was visualised at 26 Å resolution using cryo electron microscopy. At the chosen contour level the corresponding electron density maps revealed donut shaped particles with diameters of ~10 nm containing a ~20 Å wide central pore.[19,20] Later cryo-EM reconstructions with an increased resolution of ~16 Å did not show the central pore anymore but rather an indentation. Again, the number of Sec61/Y hetero-trimers fitting in the ribosome bound particles was estimated to be 3-4.[20,21]

A dimeric conformation of *E. coli* SecYEG was found when two dimensional crystals grown in liposomes were analysed with cryo EM at 8 Å resolution.[25]

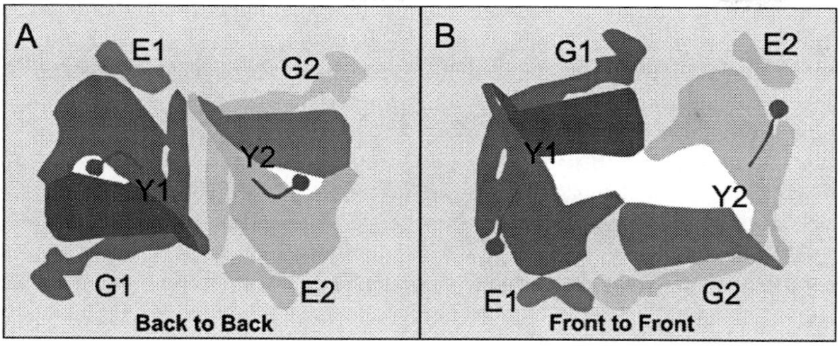

Figure 2. Cartoon representation of proposed orientations of SecYEG dimers. A) Top view of the back to back orientation as seen in 2D crystals.[23] The SecYEG monomers are coloured in shades of red and blue, respectively. Each SecYEG monomer forms a single independent pore. B) Front to front arrangement active in translocation as proposed by cryo-EM densities of the ribosome bound Sec61 complex.[33] The mouths of the monomers are facing each other forming a consolidated pore. Figure adapted from Mitra et al, 2006.[33] A color version of this image is available at www.landesbioscience.com/curie

Additional evidence that stable, nonexchangeable oligomers of SecYE form in lipid bilayers came from a fluorescence resonance energy transfer (FRET) study. FRET occured when fluorescently labelled donor and acceptor SecYE molecules were coreconstituted into liposomes showing that dimers or higher oligomers assemble.[34]

Indications about the relative orientation of the hetero-trimers in the oligomeric complex were derived by fitting the X-ray structure of SecYEβ into cryo EM maps. Two different models have been put forward. The first model predicts that a dimer of SecYEG(αβγ) with a front-to-front arrangement (the lateral gates facing each other) constitutes the active protein conducting channel.[33,35] The second model favours a back-to-back orientation with SecE forming the contact side, respectively (Fig. 2).[25,36]

To make the picture even more complicated, a recent cryo-EM and mass spectrometry study suggested that only a single SecY hetero-trimer binds to a nontranslating ribosome.[37] Also, disulfide cross-linking experiments showed that signal sequences as well as mature substrates interact only with a single SecY molecule. However, protein translocation seems to involve oligomers, because a SecY molecule defective in translocation could be rescued by cross-linking it with a wild type SecY copy.[36]

Evidently, a complete picture of the role of oligomerization cannot be depicted so far, but some tentative conclusions may be drawn:

The Sec61/Y complex active in polypeptide translocation is likely an oligomer of 2-4 hetero-trimers. Whether translocation of the polypeptide chain occurs through a hetero-trimer or a merged pore of two hetero-trimers still remains to be established. The role of additional non-translocating subunits of the complex may be the recruitment of other proteins, indispensable for protein translocation or insertion by increasing the possible contact sides between the ribosome and the Sec61/Y complex.

Structural Dynamics of the Translocon

With divergent views of the translocon structure and oligomerization as obtained from X-ray and cryo-EM studies the question arises how these data can be reconciled with different transport modes and requirements, as well as conceivable problems occurring from each of these.

It is evident that for co and posttranslational translocation, retro-translocation as well as membrane protein insertion, the pore diameter has to be wide enough to accommodate the size of helices. If, as proposed, a single Sec61/SecY hetero-trimer forms the active, transport competent channel unit,[26] the pore ring with dimensions of about 5-8 Å (center-of-atom to center-of-atom distance) has to display large flexibility with lateral displacement of the according helices. Additionally the plug domain TM2a has to be displaced to allow passage of the nascent chains. Moreover, during membrane protein integration the proposed lateral gate must be sealed against entry of lipids from the membrane into the aqueous pore.

The first all-atom molecular dynamics simulation using the SecYEβ crystal structure data showed that during pulling of deca-alanine or alanine/leucine model helices through the channel conduit with forces exceeding 1000 pN, the plug domain was destabilized and removed as proposed earlier.[38] Moreover, the pore ring opened to an inner diameter of about 7-12 Å and 6-10.5 Å, respectively. Thereby, the open channel would allow passage of complete helices. Remarkably, during the MD calculations for both model peptides the pore ring preserved a tight membrane seal since no passage of small ions was observed during translocation. After translocation of the polypeptide the structure including the plug relaxed back to its equilibrium position.

Similar results have been observed by Tian and Andricioaei.[39] In their molecular dynamics simulation where they pulled soft ball constructs with a force of 1000 pN through the SecY pore they found that with soft balls ≤9 Å, the pore ring could be expanded to diameters of about 16 Å while preserving lateral tightness against lipid influx from the membrane plane. A possible plug displacement towards the back of the channel was also observed. According relaxation simulations showed that the channel was capable of folding back to the initial shape in the time scale of nanoseconds.

In the most recent molecular dynamics simulations it was found that during lateral gating events as well as with the gate held open to the bilayer with an opening of 10-16 Å over a time interval of one microsecond, lipid molecules were prevented from invading the water-filled translocation channel by hydrophobic side chains of the gate helices TM2b and TM7 as well as the plug domain.[40] Remarkably, it was found that SecE, the proposed stabilizing clamp around SecY, had no effect on gate closure and relaxation of the structure at all.

These MD simulations show that under the applied constraints* one copy of the SecY hetero-trimer may be sufficient as the translocon conduit in terms of pore diameter required for the translocation of single trans-membrane helices and with respect to the lateral gating mechanism during membrane protein integration. However these data together with the SecY crystal structure cannot account for channel diameters necessary to accommodate disulfide-linked hairpins of preproteins or the retro-transport of almost folded proteins.[8,10] Furthermore, in experiments with fluorescent probes introduced in nascent polypeptide chains, NAD+ was found to be a potent quenching agent with full access to the nascent chain in the translocon and ribosome while Fab fragments were too large to enter the pore. From these data and the sizes of the used quenching agents, a diameter of 40-60 Å for the active channel was deduced for the Sec61 translocon.[42] These upper size dimensions were confirmed in electrophysiological experiments when the pore size of the Sec61 complex was probed with neutral polymers of different sizes. In this study two main open states were observed, one corresponding to a mean open diameter of ~2.6 nm and a second one corresponding to a mean open diameter of ~6.0 nm.[43] Additionally, for SecY locked by disulfide-bridge formation in the open state, similarly sized pore openings were observed.[31]

As outlined previously, the SecY/Sec61 hetero-trimers in the active translocon are most likely present in a higher oligomeric state.[44] However, it is still unclear how the hetero-trimers are oriented with respect to each other. The conclusions drawn from molecular dynamics simulations support the proposed "back to back"-orientation with the clamshell hinges facing each other at the subunit interface. In these orientations, the active pore is thought to be formed by only one copy of the Sec61/SecY complexes. On the contrary the front-to-front dimer assembled by two SecYEG molecules with the lateral gates facing each other opens the possibility for the formation of a single consolidated channel which could account for pore diameters of 40-60 Å as found in other experiments (see above). Moreover, this model also provides the clue for a tight seal of the aqueous translocon channel against the lipid environment of the membrane during the membrane protein insertion processes.[45]

*Very high pulling forces (≥1000 pN) were applied considering that pulling force of ≤100 pN were shown to be sufficient to unfold membrane α-helices.[41]

Conclusions

Our knowledge of the structure of the active translocon and its core, the water-filled Sec61/SecY channel has evolved drastically during the last decade. The high resolution crystal structure of the monomeric closed SecYEβ from *M. jannaschii* has been solved and the active translocon from *E. coli* and eucaryotes has been investigated with cryo-EM in several studies and with increasing resolutions. Nevertheless, all these structures are snapshots of the channel either in a monomeric or in an oligomeric state and influenced by different experimental conditions. Computational methods like molecular modelling and molecular dynamics simulations provide a useful tool to give ideas of possible structural dynamics within the translocon during the varying transport modes. However, merely a high resolution structure of the active channel with a nascent chain in transit and cryo-EM structures of the complete translocon with improved resolution combined with suitable functional studies will further extend our knowledge on the structural mechanisms of SecY/Sec61 mediated protein transport across and partition into membranes.

Acknowledgements

This work was supported by the Deutsche Forschungsgemeinschaft FOR 967.

References

1. Dalbey RE, Heijne Gv. Protein targeting, transport and translocation. In: Dalbey RE, von Heijne G, eds. Protein Targeting, Transport and Translocation. 1st ed. London: Academic Press, 2002; 1-4.
2. Simon SM, Blobel G. A protein-conducting channel in the endoplasmic reticulum. Cell 1991; 65:371-380.
3. Schatz G, Dobberstein B. Common principles of protein translocation across membranes [review]. Science 1996; 271:1519-1526.
4. Blobel G, Dobberstein B. Transfer of proteins across membranes. I. Presence of proteolytically processed and unprocessed nascent immunoglobulin light chains on membrane-bound ribosomes of murine myeloma. J Cell Biol 1975; 67:835-851.
5. Meusser B, Hirsch C, Jarosch E et al. ERAD: the long road to destruction. Nat Cell Biol 2005; 7:766-772.
6. Tsai B, Ye Y, Rapoport TA. Retro-translocation of proteins from the endoplasmic reticulum into the cytosol. Nat Rev Mol Cell Biol 2002; 3:246-255.
7. Fiebiger E, Hirsch C, Vyas JM et al. Dissection of the dislocation pathway for type I membrane proteins with a new small molecule inhibitor, eeyarestatin. Mol Biol Cell 2004; 15:1635-1646.
8. Tirosh B, Furman MH, Tortorella D et al. Protein unfolding is not a prerequisite for endoplasmic reticulum-to-cytosol dislocation. J Biol Chem 2003; 278:6664-6672.
9. Schlenstedt G, Gudmundsson GH, Boman HG et al. Structural requirements for transport of preprocecropinA and related presecretory proteins into mammalian microsomes. J Biol Chem 1992; 267:24328-24332.
10. Tani K, Tokuda H, Mizushima S. Translocation of ProOmpA possessing an intramolecular disulfide bridge into membrane vesicles of Escherichia coli. Effect of membrane energization. J Biol Chem 1990; 265:17341-17347.
11. Berridge MJ, Lipp P, Bootman MD. The versatility and universality of calcium signalling. Nat Rev Mol Cell Biol 2000; 1:11-21.
12. Schnell DJ, Hebert DN. Protein translocons: multifunctional mediators of protein translocation across membranes. Cell 2003; 112:491-505.
13. Rapoport TA, Jungnickel B, Kutay U. Protein transport across the eukaryotic endoplasmic reticulum and bacterial inner membranes. Annu Rev Biochem 1996; 65:271-303.
14. Wilkinson BM, Critchley AJ, Stirling CJ. Determination of the transmembrane topology of yeast Sec61p, an essential component of the endoplasmic reticulum translocation complex. J Biol Chem 1996; 271:25590-25597.
15. Ojakian GK, Kreibich G, Sabatini DD. Mobility of ribosomes bound to microsomal membranes. A freeze-etch and thin-section electron microscope study of the structure and fluidity of the rough endoplasmic reticulum. J Cell Biol 1977; 72:530-551.
16. Giddings TH Jr, Staehelin LA. Ribosome binding sites visualized on freeze-fractured membranes of the rough endoplasmic reticulum. J Cell Biol 1980; 85:147-152.
17. Hanein D, Matlack KE, Jungnickel B et al. Oligomeric rings of the Sec61p complex induced by ligands required for protein translocation. Cell 1996; 87:721-732.
18. Meyer TH, Menetret JF, Breitling R et al. The bacterial SecY/E translocation complex forms channel-like structures similar to those of the eukaryotic Sec61p complex. J Mol Biol 1999; 285:1789-1800.

19. Beckmann R, Bubeck D, Grassucci R et al. Alignment of conduits for the nascent polypeptide chain in the ribosome-Sec61 complex. Science 1997; 278:2123-2126.
20. Menetret JF, Neuhof A, Morgan DG et al. The structure of ribosome-channel complexes engaged in protein translocation. Mol Cell 2000; 6:1219-1232.
21. Beckmann R, Spahn CM, Eswar N et al. Architecture of the protein-conducting channel associated with the translating 80S ribosome. Cell 2001; 107:361-372.
22. Morgan DG, Menetret JF, Neuhof A et al. Structure of the mammalian ribosome-channel complex at 17A resolution. J Mol Biol 2002; 324:871-886.
23. Collinson I, Breyton C, Duong F et al. Projection structure and oligomeric properties of a bacterial core protein translocase. EMBO J 2001; 20:2462-2471.
24. Akiyama Y, Ito K. Topology analysis of the SecY protein, an integral membrane protein involved in protein export in Escherichia coli. EMBO J 1987; 6:3465-3470.
25. Breyton C, Haase W, Rapoport TA et al. Three-dimensional structure of the bacterial protein-translocation complex SecYEG. Nature 2002; 418:662-665.
26. Van den Berg B, Clemons WM Jr, Collinson I et al. X-ray structure of a protein-conducting channel. Nature 2004; 427:36-44.
27. Clemons WM Jr, Menetret JF, Akey CW et al. Structural insight into the protein translocation channel. Curr Opin Struct Biol 2004; 14:390-396.
28. Harris CR, Silhavy TJ. Mapping an interface of SecY (PrlA) and SecE (PrlG) by using synthetic phenotypes and in vivo cross-linking. J Bacteriol 1999; 181:3438-3444.
29. Tam PC, Maillard AP, Chan KK et al. Investigating the SecY plug movement at the SecYEG translocation channel. EMBO J 2005; 24:3380-3388.
30. Görlich D, Kurzchalia TV, Wiedmann M et al. Probing the molecular environment of translocating polypeptide chains by cross-linking. Methods Cell Biol 1991; 34:241-262.
31. Saparov SM, Erlandson K, Cannon K et al. Determining the conductance of the SecY protein translocation channel for small molecules. Mol Cell 2007; 26:501-509.
32. Li W, Schulman S, Boyd D et al. The plug domain of the SecY protein stabilizes the closed state of the translocation channel and maintains a membrane seal. Mol Cell 2007; 26:511-521.
33. Mitra K, Frank J. A model for cotranslational translocation: ribosome-regulated nascent polypeptide translocation at the protein-conducting channel. FEBS Lett 2006; 580:3353-3360.
34. Mori H, Tsukazaki T, Masui R et al. Fluorescence resonance energy transfer analysis of protein translocase. SecYE from Thermus thermophilus HB8 forms a constitutive oligomer in membranes. J Biol Chem 2003; 278:14257-14264.
35. Mitra K, Schaffitzel C, Shaikh T et al. Structure of the E. coli protein-conducting channel bound to a translating ribosome. Nature 2005; 438:318-324.
36. Osborne AR, Rapoport TA. Protein translocation is mediated by oligomers of the SecY complex with one SecY copy forming the channel. Cell 2007; 129:97-110.
37. Menetret JF, Schaletzky J, Clemons WM Jr et al. Ribosome binding of a single copy of the SecY complex: implications for protein translocation. Mol Cell 2007; 28:1083-1092.
38. Gumbart J, Schulten K. Molecular dynamics studies of the archaeal translocon. Biophys J 2006; 90:2356-2367.
39. Tian P, Andricioaei I. Size, motion and function of the SecY translocon revealed by molecular dynamics simulations with virtual probes. Biophys J 2006; 90:2718-2730.
40. Gumbart J, Schulten K. Structural determinants of lateral gate opening in the protein translocon. Biochemistry 2007; 46:11147-11157.
41. Janovjak H, Kessler M, Oesterhelt D et al. Unfolding pathways of native bacteriorhodopsin depend on temperature. EMBO J 2003; 22:5220-5229.
42. Hamman BD, Chen JC, Johnson EE et al. The aqueous pore through the translocon has a diameter of 40-60 A during cotranslational protein translocation at the ER membrane. Cell 1997; 89:535-544.
43. Wirth A, Jung M, Bies C et al. The Sec61p complex is a dynamic precursor activated channel. Mol Cell 2003; 12:261-268.
44. Manting EH, van Der DC, Remigy H et al. SecYEG assembles into a tetramer to form the active protein translocation channel. EMBO J 2000; 19:852-861.
45. Mitra K, Frank J. A model for cotranslational translocation: ribosome-regulated nascent polypeptide translocation at the protein-conducting channel. FEBS Lett 2006; 580:3353-3360.

CHAPTER 5

Sec Complexes in the Yeast ER

Barrie M. Wilkinson* and Colin J. Stirling*

Abstract

The transport of proteins across biological membranes is a process fundamental to intracellular protein trafficking and organelle biogenesis. Entry into the eukaryotic secretory pathway is initiated by the translocation of proteins across the membrane of the endoplasmic reticulum (ER). This occurs at the translocon, a specialised proteinaceous structure embedded in the membrane of the ER. The translocon mediates the transport of soluble proteins from the cytosol to the ER lumen and also the transfer of transmembrane domains of membrane proteins into the lipid bilayer of the ER membrane. Extensive modifications to protein structure also occur in the ER and after their biogenesis is complete, proteins are trafficked to their final subcellular destination or secreted from the cell. Whilst it has become clear that the protein conducting pore of the translocon is formed by the Sec61 complex, the precise nature of the functional translocon has yet to be ascertained. A combination of genetic, cell biology and biochemical studies in yeast has led not only to the identification of the components of the Sec61 complex, but also a number of additional components required for ER protein translocation. The roles of these components and in particular those that associate with the Sec61 complex to form the larger SEC complexes will be the focus of this chapter.

Introduction

Proteins destined for entry into the secretory pathway are targeted to the ER by distinct motifs or signals contained within the polypeptide chain. Typically, a signal sequence is located at the N-terminus and consists of a continuous stretch of hydrophobic amino acid residues (6-20) flanked by more polar amino acids (see chapter by Kapp et al for more details). After targeting has occurred, the signal sequence is usually removed in a process catalysed by the signal peptidase complex during the translocation process and the remainder of the polypeptide is translocated to the lumen of the ER. Many membrane proteins lack cleavable signal sequences and instead are targeted to the ER by way of hydrophobic signal-anchor sequences (see chapter by Cross and High for more details).

Two main pathways of targeting to the ER membrane have been described in the yeast *Saccharomyces cerevisiae*. A cotranslational pathway targets a ribosome to the ER membrane at an early stage in the translation of secretory polypeptides. This pathway in yeast has extensive similarity with that originally described in mammalian cells and crucially requires the function of the Signal Recognition Particle (SRP) and its cognate SRP Receptor (SR) in the ER membrane. In this cotranslational or SRP-dependent pathway (Fig. 1A), the ribosome nascent polypeptide complex interacts directly with the Sec61 complex and translation drives the transport of the polypeptide across the membrane. However, it has been found that some secretory polypeptides are targeted in a posttranslational or Sec62-dependent manner,[1] in which polypeptide synthesis is completed before targeting to the ER (Fig. 1B). In this pathway, proteins are maintained in an unfolded translocation competent state by interaction with cytosolic chaperones, including members of the cytosolic Ssa Hsp70 family.[2,3] However, it remains to be determined whether the mechanism of polypeptide

*Corresponding Authors: Barrie M. Wilkinson and Colin J. Stirling—Faculty of Life Sciences, University of Manchester, Manchester, M13 9PT, UK. Email: barrie.m.wilkinson@manchester.ac.uk, colin.stirling@manchester.ac.uk.

Protein Transport into the Endoplasmic Reticulum, edited by Richard Zimmermann.

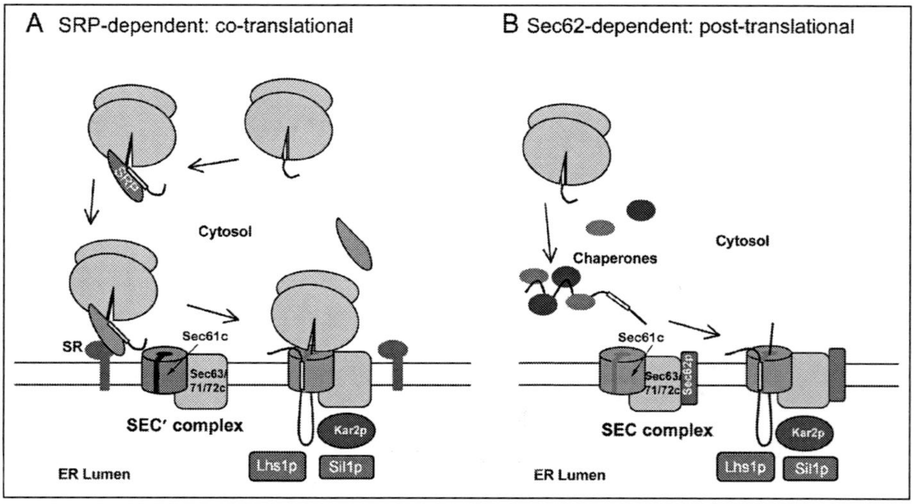

Figure 1. Endoplasmic reticulum protein translocation pathways in yeast. A) SRP-dependent (cotranslational) targeting and translocation. SRP binds to the nascent polypeptide emerging from the ribosome via the signal sequence (white) causing a delay in translation. The ribosome-SRP-nascent protein chain complex is then recruited to the ER membrane by a specific interaction between SRP and SRP receptor (SR). GTPase domains in both proteins promote GTP hydrolysis leading to the transfer of the ribosome-nascent chain complex to the translocon and the release of SRP from the ribosome and SR. The SEC′-complex consisting of the Sec61-complex (Sec61c) and the Sec63p/Sec71p/Sec72p-complex (Sec63/71/72c) forms the functional translocon. Released SRP can then facilitate further rounds of targeting whilst the ribosome resumes translation, which drives translocation of the polypeptide into the ER lumen. This mode of translocation is also completely dependent on the functions of the lumenal Hsp70, Kar2p and two nucleotide exchange factors, Lhs1p and Sil1p. B) Sec62-dependent (post-translational) targeting and translocation. A fully synthesised secretory polypeptide is bound by cytosolic chaperones maintaining a translocation competent conformation. The polypeptide is recruited to the ER membrane by virtue of the signal sequence. After signal sequence binding to the translocon which in this case is formed by the SEC-complex which crucially contains Sec62p in addition to the components of the SEC′-complex, polypeptide transfer to the ER lumen is driven by an ATP—dependent process requiring Kar2p, Lhs1p and Sil1p.

targeting in this pathway employs a signal sequence specific chaperone and a cognate receptor in the ER membrane. In both pathways, signal sequences and signal-anchor sequences directly interact with the translocon in a signal recognition step of the translocation process. Some secretory polypeptides can be targeted to the ER by either pathway, in which case they display a partial dependence on SRP. SRP dependency is determined largely by the hydrophobicity of the signal sequence, with the more lipophilic sequences being dependent on SRP.[1]

The Yeast Sec61 Complex

The pore forming component of the translocon is the evolutionarily conserved heterotrimeric membrane protein complex termed the Sec61 complex in eukaryotes and the SecYEG and SecYEβ complexes respectively in eubacteria and archaea.[4] The analysis of high resolution structures for the archaeal SecYEβ and eubacterial SecYEG complexes has identified a number of key structural features of the translocon and enabled a general model for translocon function to be proposed[4] (see chapter by Wagner et al).

The three subunits of the yeast Sec61 complex, Sec61p, Sbh1p and Sss1p show significant sequence conservation with their respective Sec61α, Sec61β and Sec61γ mammalian homologues.[5,6] The Sec61p and Sss1p subunits are both essential for cell viability and also for both the SRP- and

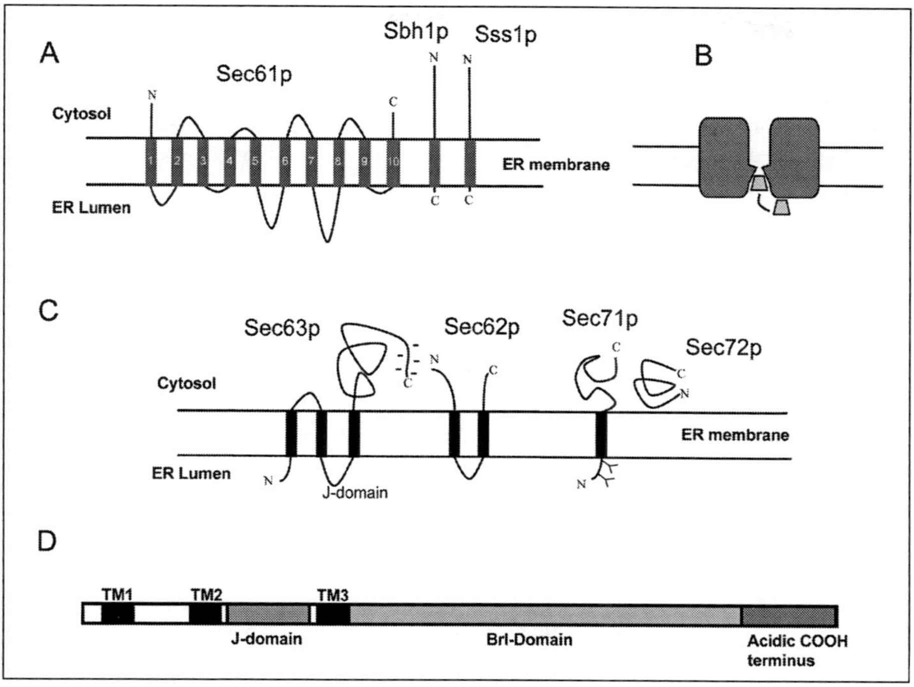

Figure 2. Membrane topology of translocon components. A) Topology of the subunits of the yeast Sec61-complex. B) A cartoon view of a translocon section formed by a single Sec61-complex, with a pore ring constriction and proposed plug domain (light grey). During translocation, this plug (a subdomain of Sec61p transmembrane domain 2) is proposed to be displaced to allow transport of the polypeptide through the pore. C) Topology of the subunits of the Sec63p complex. The Sec63p and Sec62p subunits interact via their C- and N-termini respectively. Phosphorylation of the acidic Sec63p C-terminus also plays a role in its interaction with Sec62p. The peripheral subunit Sec72p is recruited to the membrane by binding to Sec71p. D) The domain structure of Sec63p showing three transmembrane domains, a lumenal J-domain, a large cytosolic Brl-domain (371 residues) and an acidic C-terminus (52 residues) as indicated.

Sec62-dependent pathways of translocation consistent with fundamental roles in translocon function.[7,8] Sec61p is a polytopic membrane protein that spans the ER bilayer 10 times and is the major component of the protein-conducting pore (Fig. 2A).[9] A model has been proposed in which a pore formed by the Sec61p transmembrane helices contains a 'plug' which acts to open and close the lumenal face of the translocon during protein translocation (Fig. 2B).[4] The Sbh1p and Sss1p subunits are both small C-terminal anchor proteins found in tight association with Sec61p (Fig. 2A).[10] Their role in translocon function has been somewhat elusive. The Sss1 protein is proposed to interact with the hinge region of Sec61p, indicating a possible role in regulating translocon dynamics and gating.[4,6] The proposed site of Sbh1p interaction with Sec61p does not suggest any obvious role and unlike the other subunits it is not essential for cell viability and appears to play only a minor role in ER protein transport.[11,12] The roles of the Sec61 complex and its subunits are described in more detail in the chapter by Wagner et al.

Sec63p and Other Membrane Proteins Are Required for ER Protein Translocation

Genetic and biochemical studies in yeast have led to the identification of a number of other components in addition to those of the Sec61 complex, which also provide crucial functions in

ER protein translocation. These include the three integral membrane proteins Sec62p, Sec63p, Sec71p and a cytosolic peripheral membrane protein, Sec72p (Fig. 2C). Sec62p is essential for cell viability and is specifically required for posttranslational (Sec62-dependent) translocation.[13] Sec63p is also essential for cell viablity,[14] but is required for both the SRP-dependent and Sec62-dependent pathways of translocation.[15] Sec71p and Sec72p are both encoded by nonessential genes.[16,17] Although cells bearing deletions of the genes encoding Sec71p or Sec72p are specifically defective in Sec62-dependent translocation,[16,17] sec71 and sec72 mutants have been isolated that have defects in SRP-dependent translocation suggesting a possible wider role in translocation.[18] Sec72p requires Sec71p for its stable expression suggesting that Sec72p is localised to the ER membrane via interaction with the cytosolic domain of Sec71p.[19]

The Sec62p, Sec63p, Sec71p and Sec72p proteins can be isolated as a complex (the Sec63 complex) which interacts with the Sec61 complex to form a stable heptameric complex, named the SEC complex.[10,20] Proteoliposomes containing purified SEC complex can support posttranslational translocation of the secretory polypeptide prepro α factor (ppαf), upon the addition of the lumenal Hsp70 chaperone Kar2p and ATP.[10,20] The Sec61 and Sec63 complexes are clearly jointly required to form a functional posttranslational translocon in vitro and do not have any independent activity.[20] The Sec61p, Sec62p and Sec71p components interact directly with the ppαf signal sequence in the context of the SEC complex, suggesting that they associate to form a receptor for signal sequences targeted posttranslationally.[21,22] The efficient photo-crosslinking of a ppαf signal sequence simultaneously to Sec61p and Sec62p also strongly suggests that signal sequence binding occurs in a single step.[21] Whilst Sec63p does not interact with signal sequences, it performs important roles in SEC complex formation (see following section), which are crucial in bringing Sec61p, Sec62p and Sec71p into association to form the signal sequence receptor. The Sec63 protein also has a lumenal J-domain (Fig. 2D) which is not required for SEC complex formation,[23-25] but which interacts with Kar2p to propagate polypeptide translocation to the ER lumen after signal sequence binding to the translocon (see later section on Hsp70 function). Hence, Sec63p has roles at two distinct steps of a Sec62-dependent translocation reaction; signal sequence recognition and Kar2p/ATP-dependent translocation of a polypeptide to the ER lumen.

The SEC Complexes

Structural studies suggest that the Sec61 complex is a heterotrimer containing one molecule each of Sec61p, Sss1p and Sbh1p.[4] However, the overall stoichiometry of the SEC complex and how its various components interact is only partially resolved. A specific interaction between Sec63p and Sec62p has been well characterized. The N-terminal cytosolic domain of Sec62p interacts directly with last 14 amino acid residues of Sec63p.[26,27] Yeast cells expressing an allele of Sec63p lacking these last 14 residues are viable and have defects specifically in Sec62-dependent translocation.[24,27] This interaction between Sec62p and Sec63p is required to maximize the efficiency of Sec62-dependent translocation, since it is not essential for this process. The reduced level of Sec62-dependent translocation that occurs in its absence may be explained by a much lower affinity interaction of Sec62p with the SEC complex mediated via its cytosolic C-terminus.[26]

The native gel analysis of Sec61p and Sec63p containing complexes derived from solubilised yeast ER membranes has revealed an intriguing scenario of five complexes.[23] Four Sec61p containing complexes migrating at approx. 140, 280, 350 and 380 kDa were detected. The 380 kDa complex appears to be the equivalent of the SEC complex as it is the only one containing all seven components of this heptameric complex. The 350 kDa SEC′ complex contains all the components of the SEC complex, but interestingly lacks Sec62p. The 140 and 280 kDa complexes contain Sec61p, Sss1p and Sbh1p and by analogy with the behaviour of SecYEG complexes on native gels may represent monomeric and dimeric forms of the Sec61-complex respectively.[28] The tetrameric Sec63p complex which is usually isolated by Sec62p affinity methods,[10,20] was not detected in this analysis, but a complex containing Sec63p, Sec71p and Sec72p migrating at approx 220 kDa was clearly found.

Based on the known functions of the components of the SEC complex, it is reasonable to speculate that the SEC and SEC′ complexes represent the posttranslational (Sec62-dependent) and cotranslational (SRP-dependent) translocons respectively. This is supported by a number of findings. First when cells are depleted of Sec62p, the SEC complex is lost and there is a concomitant block in posttranslational translocation.[23] Second when cells are depleted of Sec63p, both the SEC and SEC′ complexes are lost and both translocation pathways are blocked. Under both of these depletion conditions, the 140/280 kDa Sec61-containing complexes remain abundant, strongly suggesting that these smaller complexes are themselves not sufficient to support translocation. Whilst the pools of the 140/280 kDa sec61- and the Sec63/Sec71/Sec72 complexes may result from SEC′ instability under experimental conditions, they may represent bonafide intermediates in a pathway of translocon assembly (see Fig. 3). The presence of Sec71p and Sec72p in the cotranslational SEC′ complex may reflect a possible role in this pathway as previously indicated.[18]

The higher order SEC and SEC′ complexes also provide the context in which Sec63p functions during ER protein translocation. The cytosolic domain of Sec63p contains two subdomains which are required for the formation of the SEC and SEC′ complexes. The extreme C-terminal 52 residues of Sec63p are predominantly acidic and the last 14 of these have been shown to interact directly with Sec62p consistent with their requirement to form the SEC complex and hence the posttranslational translocon.[23] Moreover, the phosphorylation of specific residues within the acidic domain of Sec63p contribute to Sec62p binding suggesting a mechanism to promote Sec62p binding and regulate the

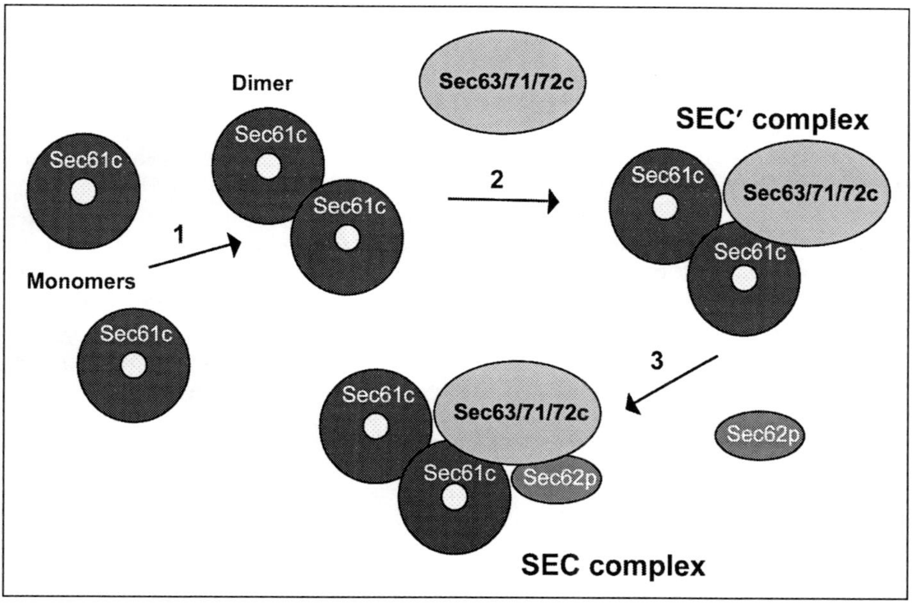

Figure 3. A proposed assembly pathway for the SEC and SEC′ complexes. The yeast ER membrane is proposed to contain a number of complexes capable of assembly into the higher order SEC and SEC′ complexes. The scheme shown here (looking down through membrane from cytosol) is the most economical interpretation based on the sizes of complexes on native gels. In step 1 two monomers of Sec61-complexes (Sec61c) dimerise. The SEC′ complex would then be formed in step 2 by the stable interaction between a dimer of Sec61 complexes and the Sec63p/Sec71p/Sec72p complex (Sec63/71/72c). The larger SEC complex would in turn be formed by recruitment of Sec62p in step 3 which may be regulated by the phosphorylation status of Sec63p, thus converting a SEC′ complex into a SEC complex. Although a number of studies indicate that functional translocons contain two Sec61 complexes as shown here, only one of these appears to form the functional pore (see ref. 4).

relative levels of SEC and SEC′ complexes within the ER membrane.[29] Immediately adjacent to the acidic subdomain of Sec63p is a conserved Brl domain consisting of 371 residues (Fig. 2D) which is required for the assembly of both the SEC and SEC′ complexes.[23] This has led to the conclusion that the cytosolic domain of Sec63p performs two distinct roles in translocon assembly. The Brl subdomain is required for the formation of the SEC′ complex and the acidic C-terminal domain is then required to recruit Sec62p to form the larger SEC complex. The recruitment of Sec62p may not only convert a cotranslational translocon into a posttranslational translocon, but may conceivably exclude it from performing cotranslational translocation. How the Sec61- and Sec63- containing complexes interact to form the higher order complexes is not clear, but it may be mediated by a direct interaction between Sec61p and Sec63p as suggested by genetic data.[30] The SEC and SEC′ complexes may provide a locus to recruit chaperones/cochaperones to both the cytosolic and lumenal faces of the translocon and also to recruit other additional components essential for ER protein biogenesis including the signal peptidase and oligosaccharyl transferase complexes.

The Role of Hsp70 Function in ER Protein Translocation

The translocation of polypeptides into the ER lumen requires not only the membrane embedded SEC and SEC′ complexes, but is also critically dependent on several lumenal components. A canonical member of the Hsp70 family of chaperones, Kar2p, which is the yeast equivalent of mammalian BiP is essential for both pathways of translocation.[15,31] Members of the Hsp70 class of chaperones play essential roles in a number of diverse cellular processes including the folding and assembly of newly synthesized proteins and the transport of proteins across membranes. Hsp70 chaperones interact directly with polypeptides in a nucleotide-dependent manner and undergo conformational changes in a cycle requiring a J-protein to trigger ATP hydrolysis and enhance substrate binding and a nucleotide exchange factor (NEF) to remove ADP (Fig. 4A, also see chapter by Dudek et al).[32] There is compelling evidence that the lumenal J-domain of Sec63p binds Kar2p in an ATP-dependent manner.[33-35] This J-domain is essential for cell viability and for both translocation pathways,[36] but is not required for the assembly of the SEC′ and SEC complexes.[23] The Kar2 protein is not required for polypeptide targeting to the SEC complex, but is required to propagate translocation.[20] This strongly suggests that the Sec63p J-domain and Kar2p together perform a role which is distinct from the roles of the cytosolic domain of Sec63p. The nucleotide exchange activity is provided by two Kar2p binding proteins, Lhs1p, a member of the Hsp110 chaperone family and Sil1p, a member of the BAG domain proteins.[37-40] Neither Lhs1p nor Sil1p is required for cell viability, but cells lacking both proteins are inviable with a corresponding complete block of protein transport into the ER.[39] Thus, yeast clearly has the three components required to support an ER Hsp70 cycle localized to the lumenal face of the translocon by the Sec63p J-domain. The fact that all three components are essential for ER protein translocation irrespective of the pathway suggests that they are jointly performing a fundamental role(s) in the process.

Two different roles can explain the function of an Hsp70 cycle during ER protein translocation. The protein conducting pore of the translocon is highly likely to be tightly regulated to maintain the permeability barrier of the ER membrane. Although, the high resolution structure of the translocon shows a plug domain that is predicted to seal the closed translocon (see Fig. 2B and chapter by Wagner et al), in mammalian cells, BiP is required to gate the lumenal face of the translocon. A model in which BiP regulates translocon pore closure and opening by interacting with a membrane bound J-domain and also another unidentified region of the translocon has been proposed.[41] The translocon is sealed by the ADP-bound conformation of BiP and pore opening requires an ATP binding-induced conformational change requiring NEF activity to displace the ADP. Resealing would then occur when BiP interacts with translocon followed by J-domain stimulated ATP hydrolysis. The opening and sealing of the translocon pore at the lumen would be consistent with a fundamental role played by Kar2p, Sec63p and Lhs1p/Sil1p in both translocation pathways (Fig. 4B).

However, it also seems likely that these proteins would also play another essential role in providing directionality during posttranslational translocation (Fig. 1A). The sequential binding of

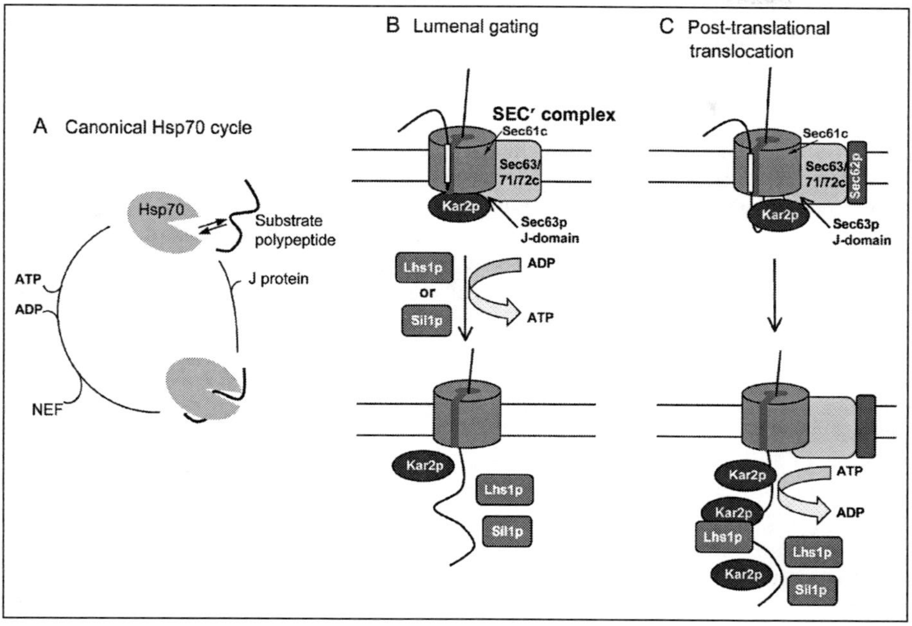

Figure 4. Proposed roles for Hsp70 function in ER protein translocation. A) Schematic of a canonical Hsp70 cycle. The Hsp70 interacts directly with a polypeptide and undergoes comformational changes. A J-protein triggers ATP-hydrolysis and enhances substrate binding, whilst a nucleotide exchange factor (NEF) stimulates ADP release, allowing ATP to bind before a new cycle of substrate interaction. B) Gating of the lumenal face of the SEC and SEC′ translocon (as shown) in both the SRP- and Sec62-dependent pathways. ADP-bound Kar2p is bound to the Sec63p J-domain and another unidentified part of the translocon which acts as the substrate. In order for translocation to proceed, Kar2p is displaced when ADP is removed by one of the NEFs (Lhs1p or Sil1p) and exchanged for ATP thus opening the translocon to the ER lumen. ATP hydrolysis upon Sec63p J-domain interaction would then be required for Kar2p rebinding to the translocon. C) Post-translational (Sec62-dependent) translocation. In this model, Kar2p binds the J-domain transiently to trigger ATP hydrolysis and facilitate its binding to the incoming polypeptide substrate. The polypeptide is prevented from diffusing back to the cytosol by the binding of multiple Kar2p molecules in a 'ratchet' mechanism. The NEF activity of Lhs1p or Sil1p would then be required to remove ADP from Kar2p, allowing Kar2p to disassociate from the polypeptide, allowing further cycles of Sec63p J-domain and ATP-dependent Kar2p binding to the polypeptide. Lhs1p may also contribute to this mechanism by directly binding polypeptides in the context of Kar2p/Lhs1p heterodimers.

Kar2p molecules with the incoming polypeptide in a 'ratchet' mechanism mediated by the Sec63p J-domain, may provide this directionality (Fig. 4C).[42] The NEF activity of Lhs1p/Sil1p is then predicted to promote the disassociation of Kar2p from the polypeptide allowing Kar2p to embark on a new round of Sec63p and ATP stimulated binding of polypeptide. Studies of other Hsp110 molecules suggest that Lhs1p may have a more complex role. A structural analysis has shown that cytosolic Hsp110 and Hsp70 molecules form heterodimers with their substrate binding domains in close apposition. This has led to a model where both subunits interact with different portions of the substrate and with the Hsp110 regulating the nucleotide-dependent cycling of the Hsp70 by allosteric coupling.[43] Given that it has been shown that a mammalian homologue of Lhs1p has polypeptide binding activity, this raises the possibility that Kar2p and Lhs1p also form heterodimers in the yeast ER which co-operatively bind incoming polypeptides during posttranslational translocation (Fig. 4C).[44]

A 'molecular motor' model in which Kar2p/NEF would be bound to the translocon via the Sec63p J-domain performing rapid rounds of polypeptide binding and release to drive the polypeptide into the lumen via Kar2p conformational changes also remains a possibility. There is considerable evidence for a motor complex at the yeast mitochondrial preprotein translocase consisting of an Hsp70, a NEF, a membrane bound J protein and a number of other J-like proteins.[45] Interestingly, the protein biogenesis factors recruited to the polypeptide exit site of the ribosome including two Hsp70 homologues, an Hsp110, a BAG domain NEF and a J-protein (see ref. 46 and references therein) bear considerable similarity with those components required for protein translocation into the ER lumen and the mitochondrial matrix. This raises the possibility that a similar complex of chaperones and cochaperones may function at the ER lumenal face of the translocon and that such complexes may be a general feature of regulating the biogenesis of newly synthesized proteins.

Concluding Remarks

Whilst it is established that the protein conducting pore of the translocon is formed by the conserved Sec61 complex, it has become increasingly apparent that translocation across the ER membrane of yeast is dependent on the function of many other components. The essential components, Sec63p, Sec62p, Kar2p and Lhs1p/Sil1p all have conserved counterparts in mammalian cells, suggesting that translocation is also mechanistically conserved. The composition of the translocon differs depending on whether a secretory polypeptide is undergoing co or posttranslational translocation as represented by the SEC and SEC' complexes. The Sec63 protein performs essential roles in the assembly of these complexes and also has a distinct role in recruiting Kar2p to the translocon, thus enabling protein translocation to be coupled to an Hsp70 cycle. Given these roles in translocon function, there is a possibility that Sec63p acts to regulate the translocon and co-ordinate different steps of the translocation reaction. Now that the key components required for ER protein transport have been defined, a number of challenging questions remain. The determination of the exact stoichiometric composition of functioning translocon complexes and also a resolution as to whether the translocon is gated by Sec61p or BiP/Kar2p are among the challenges. Furthermore, a more detailed analysis of the precise role(s) played by an Hsp70 cycle and whether a stable J-protein/Hsp70/NEF complex functions at the lumenal side of the translocon also await further developments.

Acknowledgements

Research in the lab of C.J.S. is supported by funding from the BBSRC, the MRC and the Wellcome Trust. We wish to thank Jeanine de Keyzer and Sarah Mahoney for helpful comments during the preparation of this chapter.

References

1. Ng DTW, Brown JD, Walter P. Signal sequences specify the targeting route to the endoplasmic reticulum membrane. J Cell Biol 1996; 134:269-278.
2. Wilkinson BM, Regnacq M, Stirling CJ. Protein translocation across the membrane of the endoplasmic reticulum. J Membr Biol 1997; 155:189-197.
3. Plath K, Rapoport TA. Release of cytosolic proteins from posttranslational substrates before their transport into the endoplasmic reticulum. J Cell Biol 2000; 151:167-178.
4. Osborne AR, Rapoport TA, van den Berg B. Translocation by the sec61/SECY channel. Annu Rev Cell Dev Biol 2004; 21:529-550.
5. Hartmann E, Sommer T, Prehn S et al. Evolutionary conservation of components of the protein translocation complex. Nature 1994; 367:654-657.
6. Van den Berg B, Clemons WM Jr, Collinson I et al. X-ray structure of a protein-conducting channel. Nature 2004; 427:36-44.
7. Stirling CJ, Rothblatt J, Hosobuchi M et al. Protein translocation mutants defective in the insertion of integral membrane proteins into the endoplasmic reticulum. Mol Biol Cell 1992; 3:129-142.
8. Esnault Y, Blondel MO, Deshaies RJ et al. The yeast SSS1 gene is essential for secretory protein translocation and encodes a conserved protein of the endoplasmic-reticulum. EMBO J 1993; 12:4083-4093.

9. Wilkinson BM, Critchley AJ, Stirling CJ. Determination of the transmembrane topology of yeast sec61p, an essential component of the endoplasmic reticulum translocation complex. J Biol Chem 1996; 271:25590-25597.
10. Panzner S, Dreier L, Hartmann E et al. Posttranslational protein transport in yeast reconstituted with a purified complex of sec proteins and kar2p. Cell 1995; 81:561-570.
11. Finke K, Plath K, Panzner S et al. A second trimeric complex containing homologs of the sec61p complex functions in protein-transport across the ER membrane of S. cerevisiae. EMBO J 1996; 15:1482-1494.
12. Feng D, Zhao X, Soromani C et al. The transmembrane domain is sufficient for Sbh1p function, its association with the sec61 complex and interaction with rtn1p. J Biol Chem 2007; 282:30618-30628.
13. Deshaies RJ, Schekman R. SEC62 encodes a putative membrane protein required for protein translocation into the yeast endoplasmic reticulum. J Cell Biol 1989; 109:2653-2664.
14. Sadler I, Chiang A, Kurihara T et al. A yeast gene important for the assembly into the endoplasmic reticulum and the nucleus has homology to DnaJ, an Escherichia coli heat shock protein. J Cell Biol 1989; 109:2665-2675.
15. Young BP, Craven RA, Reid PJ et al. Sec63p and kar2p are required for the translocation of SRP-dependent precursors into the yeast endoplasmic reticulum in vivo. EMBO J 2001; 20:262-271.
16. Feldheim D, Yoshimura K, Admon A et al. Structural and functional characterization of sec66p, a new subunit of the polypeptide translocation apparatus in the yeast endoplasmic reticulum. Mol Biol Cell 1993; 4:931-939.
17. Feldheim D, Sckuekman R. Sec72p contributes to the selective recognition of signal peptides by the secretory polypeptide translocation complex. J Cell Biol 1994; 126:935-943.
18. Green N, Fang H, Walter P. Mutants in three novel complementation groups inhibit membrane protein insertion into and soluble protein translocation across the endoplasmic reticulum membrane of saccharomyces cerevisiae. J Cell Biol 1992; 126:597-604.
19. Fang H, Green N. Nonlethal sec71-1 and sec72-1 mutations eliminate proteins associated with the sec63p-BiP complex from S. cerevisiae. Mol Biol Cell 1994; 5:933-942.
20. Matlack KE, Plath K, Misselwitz B et al. Protein transport by purified yeast sec complex and kar2p without membranes. Science 1997; 277:938-941.
21. Plath K, Mothes W, Wilkinson BM et al. Signal sequence recognition in posttranslational protein transport across the yeast ER membrane. Cell 1998; 94:795-807.
22. Plath K, Wilkinson BM, Stirling CJ et al. Interactions between sec complex and prepro-alpha-factor during posttranslational protein transport into the endoplasmic reticulum. Mol Biol Cell 2004; 15:1-10.
23. Jermy AJ, Willer M, Davies E et al. The brl domain in sec63p is required for assembly of functional endoplasmic reticulum translocons. J Biol Chem 2006; 281:7899-906.
24. Feldheim D, Rothblatt J, Schekman R. Topology and functional domains of sec63p, an endoplasmic reticulum membrane protein required for secretory protein translocation. Mol Biol Cell 1993; 12:3288-3296.
25. Lyman SK, Schekman R. Interaction between BiP and sec63p is required for the completion of protein translocation into the ER of saccharomyces cerevisiae. J Cell Biol 1995; 131:1163-1171.
26. Wittke S, Dunnwald M, Johnsson N. Sec62p, a component of the endoplasmic reticulum protein translocation machinery, contains multiple binding sites for the sec-complex. Mol Biol Cell 2000; 11:3859-3871.
27. Willer M, Jermy AJ, Young BP et al. Identification of novel protein-protein interactions at the cytosolic surface of the sec63 complex in the yeast ER membrane. Yeast 2003; 20:133-148.
28. Duong F. Binding, activation and dissociation of the dimeric SecA ATPase at the dimeric SecYEG translocase. EMBO J 2002; 22:4375-4384.
29. Wang X, Johnsson N. Protein kinase CK2 phosphorylates sec63p to stimulate the assembly of the endoplasmic reticulum protein translocation apparatus. J Cell Sci 2005; 118:723-732.
30. Pilon M, Römisch K, Quach D et al. Sec61p serves multiple roles in secretory precursor binding and translocation into the endoplasmic reticulum membrane. Mol Biol Cell 1998; 9:3455-3473.
31. Vogel JP, Misra LM, Rose MD. Loss of BiP/GRP78 function blocks translocation of secretory proteins in yeast. J Cell Biol 1990; 110:1885-1895.
32. Bukau B, Weissman J, Horwich A. Molecular chaperones and protein quality control. Cell 2006; 125:443-451.
33. Corsi AK, Schekman R. The lumenal domain of sec63p stimulates the ATPase activity of BiP and mediates BiP recruitment to the translocon in saccharomyces cerevisiae. J Biol Chem 1997; 137:1483-1493.
34. Misselwitz B, Staeck O, Matlack KE et al. Interaction of BiP with the J-domain of the sec63p component of the endoplasmic reticulum protein translocation complex. J Biol Chem 1999; 274:20110-20115.

35. McLellan AJ, Endres JB, Vogel JP et al. Specific molecular chaperone interactions and an ATP-dependent conformational change are required during posttranslational protein translocation into the yeast ER. Mol Biol Cell 1998; 9:3533-3545.
36. Willer M, Jermy AJ, Steel GJ et al. An in vitro assay using overexpressed yeast SRP demonstrates that cotranslational translocation is dependent upon the J-domain of sec63p. Biochemistry 2003; 42:7171-7177.
37. Craven RA, Egerton M, Stirling CJ. A novel Hsp70 of the yeast ER lumen is required for the efficient translocation of a number of protein precursors. EMBO J 1996; 15:2640-2650.
38. Steel GJ, Fullerton DM, Tyson JR et al. Coordinated activation of Hsp70 chaperones. Science 2002; 303:98-101.
39. Tyson JR, Stirling CJ. LHS1 and SIL1 provide a lumenal function that is essential for protein translocation into the endoplasmic reticulum. EMBO J 2000; 19:6440-6452.
40. Kabani M, Beckerich KM, Gaillardin C. Sls1p stimulates sec63p-mediated activation of kar2p in a conformation-dependent manner in the yeast endoplasmic reticulum. Mol Cell Biol 2000; 20:6923-6934.
41. Alder NN, Shen Y, Brodsky JL et al. The molecular mechanisms underlying BiP-mediated gating of the sec61 translocon of the endoplasmic reticulum. J Cell Biol 2005; 168:389-399.
42. Matlack KE, Misselwitz B, Plath K et al. BiP acts as a molecular ratchet during posttranslational transport of prepro-alpha factor across the ER membrane. Cell 1999; 97:553-564.
43. Polier S, Dragovic Z, Hartl U et al. Structural basis for the cooperation of Hsp70 and Hsp110 chaperones in protein folding. Cell 2008; 133:1068-1079.
44. Park J, Easton DP, Chen X et al. The chaperoning properties of mouse grp170, a member of the third family of hsp70 related proteins. Biochemistry 2003; 42:14893-14902.
45. Bohnert M, Pfanner N, van der Laan M. A dynamic machinery for import of mitochondrial precursor proteins. FEBS Lett 2007; 581:2802-2810.
46. Conz C, Otto H, Peisker K et al. Functional characterization of the atypical hsp70 subunit of yeast ribosome-associated complex. J Biol Chem 2007; 282:33977-33984.

Chapter 6

The Role of BiP and Its Co-Chaperones

Johanna Dudek,* Julia Benedix and Carolin Jalal

Abstract

Molecular chaperones and their co-chaperones of the endoplasmic reticulum (ER) are crucial for efficient ER-associated protein biogenesis. The major molecular chaperone of the ER, BiP, is a member of the heat shock protein 70 (Hsp70) family and cooperates with co-chaperones of the Hsp40 family, as well as with nucleotide exchange factors. So far, six Hsp40 co-chaperones, ERj1 to ERj6 and two nucleotide exchange factors, Sil1 and Grp170, have been identified in mammals. In addition to the classical role of Hsp40 co-chaperones in stimulating the ATPase activity of BiP via the characteristic J-domain, ERj proteins mediate various specific BiP functions based on their additional individual structural and functional properties. During protein transport into the ER, BiP cooperates with the two membrane-anchored ERj proteins, ERj1 and ERj2/Sec63.

Introduction

Protein import into the endoplasmic reticulum (ER) can occur co- or posttranslationally and involves amino terminal signal peptides in the precursor proteins as well as a protein translocase that has the Sec61 complex as its central component.[1] After transport into the lumen of the ER or integration into the ER membrane, the proteins reach their native conformation. This involves posttranslational modifications, folding of the polypeptide chain and in case of heteromeric proteins assembly into protein complexes. Only in the native conformation, proteins are transported further to their final destination by vesicular transport. If a protein cannot achieve its native conformation due to mutations or cellular stress, it is exported to the cytosol for removal by the proteasome in a process called ER-associated degradation (ERAD).[2,3]

Molecular chaperones of the Hsp70 family and their Hsp40-type co-chaperones are required for all steps of ER-associated protein biogenesis: translocation into the ER, folding and assembly in the lumen and export of misfolded polypeptides out of the ER followed by proteasomal degradation.[4] If the balance among these processes is disturbed, for instance during stress conditions, the "unfolded protein response" (UPR) is activated; chaperones are also involved in this process. They serve as sensors for the accumulation of misfolded proteins and initiate signal transduction out of the ER. BiP is the best-characterized ER Hsp70 chaperone involved in these processes. BiP interacts directly with UPR mediators in the ER membrane and maintains them in their inactive form. When misfolded proteins accumulate, BiP is needed for folding and dissociates from UPR mediators, initiating the transduction of the unfolded protein signals across the ER membrane to the cytosol and to the nucleus. The UPR includes two ways of modulation: the induction of global translational arrest and transcriptional regulation. Translation arrest prevents further overloading of the ER lumen with folding substrates, whereas transcriptional regulation leads to increased synthesis of ER-resident chaperones, which assist in proper folding of accumulated polypeptides.[5,6]

*Corresponding Author: Johanna Dudek—Medizinische Biochemie und Molekularbiologie, Universität des Saarlandes, D-66421 Homburg, Germany. Email: bcjdud@uks.eu

Protein Transport into the Endoplasmic Reticulum, edited by Richard Zimmermann

The Hsp70/Hsp40 Chaperone System

Hsp70 chaperones are highly conserved proteins of about 70 kDa and are found in all organisms.[7] All Hsp70s are composed of three domains, an N-terminal ATPase domain, a substrate-binding domain and a carboxy-terminal lid domain.[7-9] Hsp70s function by binding and releasing hydrophobic regions of their substrate, which usually are buried inside native proteins but are unprotected in unfolded polypeptides. By protecting the hydrophobic regions, Hsp70s give their substrates time to achieve their native state.[10] This process is ATP-dependent and for most Hsp70s, ATP hydrolysis is the rate-limiting step.[11,12] The Hsp70 ATPase cycle is regulated by co-chaperones, such as proteins of the Hsp40 family and nucleotide exchange factors. Hsp40s stimulate the weak intrinsic ATPase activity of Hsp70s, whereas nucleotide exchange factors cause the release of ADP (Fig. 1).[13-16]

The defining feature of the heterogeneous Hsp40 protein family is the highly conserved J-domain. The J-domain is required both for the ATP-dependent interaction of the Hsp40 co-chaperone with its Hsp70 partner and for the stimulation of the Hsp70's ATPase activity.[17,18] The J-domain comprises approximately 70 amino acids and is typically located at the amino terminus of the Hsp40s.[19,20] Maximal stimulation of the ATPase activity by the J-domain is achieved in the presence of a folding substrate.[21] Some Hsp40 proteins bind directly to polypeptide chains and

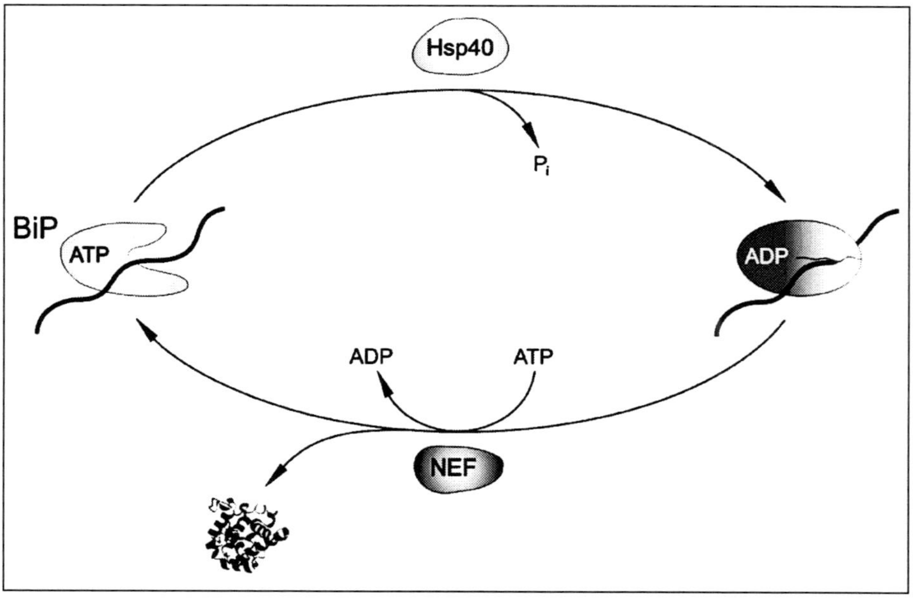

Figure 1. The ATPase cycle of BiP. The folding substrates are either recognized by BiP, then an Hsp40 co-chaperone enters the cycle, or are presented BiP by an Hsp40 protein. Owing to their additional domains, the Hsp40s ERj3, ERj4 ERj5 and ERj6 may be able to bind polypeptide substrates and deliver them to the peptide-binding domain of BiP. In the case of ERj1 and ERj2, BiP appears to be recruited to the substrate polypeptides by spatial proximity to the Sec61 complex and ribosomes, respectively. In the ATP-bound state, BiP has a low affinity for substrate polypeptides. Upon binding of the J-domain of an Hsp40 co-chaperone to the ATPase domain of BiP, ATP is hydrolyzed and a conformational change is triggered in the peptide binding domain of BiP hence "closing" BiP onto the unfolded polypeptide. Nucleotide exchange stimulated by a nucleotide exchange factor such as Grp170 in the pancreatic ER reverses the conformational shift in BiP and releases the folding substrate, allowing it to fold or to re-enter the cycle. We believe that a transient substrate-like interaction of the J-domain with the peptide binding domain of BiP may also be a part of the cycle.
Abbreviations: NEF: nucleotide exchange factor; P_i: inorganic phosphate.

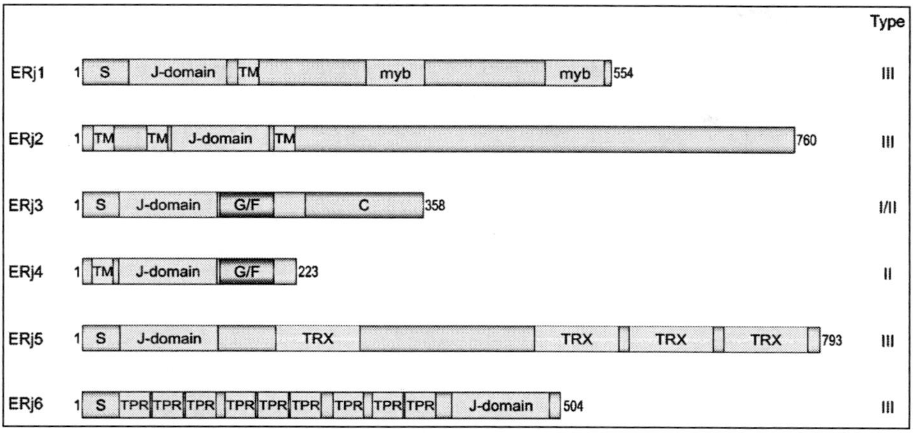

Figure 2. Schematic domain organization of the Hsp40 co-chaperones of the mammalian ER. The classification of the Hsp40 type is indicated on the right. The protein ERj3 contains a cysteine-rich domain instead of the canonical four cysteine repeats. Therefore, it cannot be clearly assigned to the group of type I Hsp40 proteins.[84]
Abbreviations: C: cysteine-rich domain; G/F: glycine/phenylalanine-rich region; myb: Myb domain; TM: transmembrane domain; TPR: tetratricopeptide repeat; TRX: thioredoxin domain; S: signal peptide.

protect them from aggregation or transfer them to an Hsp70 protein.[22,23] Based on their structural properties, Hsp40s are divided into three classes (Fig. 2).Type I Hsp40s are most homologous to the archetypical *E. coli* chaperone, DnaJ. The J-domain is followed by a glycine/phenylalanine-rich region, four cysteine repeats and a nonconserved carboxy terminus. Type II Hsp40s lack the cysteine repeats and Type III proteins contain only the conserved J-domain.[24] Most Hsp40 proteins contain additional structural elements that enable the Hsp70/Hsp40 system to fulfill a variety of specific cellular functions.

In the presence of an Hsp40 co-chaperone, the rate limiting step in the Hsp70 ATPase cycle becomes the release of ADP.[25] A heterogeneous group of proteins, the nucleotide exchange factors, stimulate the ATPase activity by enhancing the release of ADP from Hsp70s.[16,26-31]

In a typical Hsp70/Hsp40 cycle (Fig. 1), an Hsp70 binds to an unfolded polypeptide; the peptide-binding pocket is open and substrate binding is weak. Stimulation by an Hsp40 triggers strong binding to the polypeptide. Substrate release occurs after ADP/ATP exchange catalyzed by a nucleotide exchange factor. If the polypeptide has not reached its native conformation, it will bind again and a further cycle is initiated.[32]

Hsp70/Hsp40 Chaperones of the ER

There are two members of the Hsp70 family in theER, Kar2p and Lhs1p in yeast and their orthologues BiP and Grp170 in mammals (Table 1).[33-36] Both, Kar2p and BiP are essential and UPR-regulated proteins.[37-40] BiP constitutes the major ER-lumenal chaperone and plays a role in a variety of ER functions including protein import into the ER, gating the protein-conducting translocase, folding nascent proteins, targeting misfolded proteins for degradation by ERAD, regulating the UPR and contributing to regulation of ER calcium stores.[36,41-45] Except for calcium binding, all of these functions require the ATPase activity of BiP, which is regulated by co-chaperones of the Hsp40 family and nucleotide exchange factors. The diverse functions of BiP are specified by the different ER-resident Hsp40 proteins (see below).

Grp170/Lhs1p, an atypical Hsp70 protein related to Hsp110 chaperone family members, operates as a nucleotide exchange factor for BiP.[46-48] Grp170 forms a stable complex with BiP in the absence of ATP and is the major nucleotide exchange factor for BiP in pancreatic cells (Table 2).[47]

Table 1. Hsp70 chaperones und their co-chaperones in the ER of yeast Saccharomyces cerevisiae and mammals. The concentrations refer to a suspension of canine pancreatic rough microsomes (RM), with a concentration of the heterotrimeric Sec61 complexes of 2.12 μM.[73] n.d., not determined.

Protein Family	Yeast Protein (Alternative Name)	Mammalian Protein (Alternative Name)	Mammalian Concentration in RMS (μM)
Hsp70	Kar2p (Grp78)	BiP (Grp78, Hspa5)	5.00
	Lhs1p (Cer1p, Ssi1p)	Grp170 (Hyou1, Orp150)	0.60
Hsp40	-	ERj1 (Mtj1, ERdj1, DnaJC1)	0.36
	Sec63p (Ptl1p, Npl1p)	ERj2 (Sec63)	1.98
	Scj1p	ERj3 (ERdj3, HEDJ, Dj9, DnaJB11)	0.29
	-	ERj4 (ERdj4, MDG1, DnaJB9)	0
	-	ERj5 (ERdj5, JPDI, DnaJC10)	2.00
	-	ERj6 (p58^{IPK}, DnaJC3)	n.d.
	Erj5p (YFR041C, Scj2p)	-	
	Jem1p	-	
NEF	Sil1p (Sls1p)	Sil1 (BAP)	0.005

The second nucleotide exchange factor in the ER is Sil1/Sil1p. Both *LHS1* and *SIL1* are nonessential genes in yeast. However, simultaneous deletion of both genes results in synthetic lethality.[28] Mutations in the human *SIL1* gene are linked to the neurodegenerative disease MarinescoSjögren syndrome (see chapter by Müller et al).[48,49]

In the yeast *Saccharomyces cerevisiae,* four ER Hsp40 proteins containing a lumenal J-domain have been identified: the two membrane proteins Sec63p and Erj5p and the two soluble lumenal proteins Scj1p and Jem1p (Table 1).[50-54] The essential Sec63p is part of the heptameric Sec complex

Table 2. The affinities of mammalian Hsp40 co-chaperones for BiP and stimulation of the ATPase activity by ERj proteins and nucleotide exchange factors. n.d., not determined.

Protein	Affinity for BiP in the Presence of ATP (K_D in μM)	Stimulation of ATPase Activity of BiP (Fold)	Further Stimulation of BiP ATPase By Grp170 (Fold)	Further Stimulation of BiP ATPase By Sil1 (Fold)
ERj1	0.12	5.2	5.6	2.4
ERj2	5.00	3.5	5.7	2.2
ERj3	3.60	1.8	1.9	1.3
ERj4	6.07	1.6	1.5	1.0
ERj5	0.45	3.9	4.3	1.1
ERj6	n.d.	n.d.	n.d.	n.d.

that mediates posttranslational protein transport into or across the ER membrane.[50,55] Together, the yeast BiP homologue Kar2p and Sec63p are involved in all stages of posttranslational translocation.[56-59] It has been reported that Sec63p and Kar2p also function in cotranslational protein transport and retrograde transport of misfolded proteins from the ER to the cytosol.[60-63] Scj1p and Jem1p are regulated by UPR. These nonessential proteins are involved in Kar2p functions necessary for protein folding and ERAD.[64,65] Loss of the nonessential UPR-regulated *ERJ5* gene leads to constitutively induced UPR, indicating that Erj5p is required to preserve the folding capacity of the yeast ER.[52]

Six Hsp40 proteins have been identified in the ER of mammalian cells, ERj1/Mtj1, ERj2/Sec63, ERj3, ERj4, ERj5 and ERj6/p58^{IPK} (Fig. 2).[66-71] These proteins have been found in varying amounts in canine pancreatic rough microsomes (Table 1). The J-domains of all of these co-chaperones interact with BiP but their affinities for BiP differ (Table 2). The amounts of Hsp70 and Hsp40 proteins present in the ER lumen (Table 1) and their affinities for BiP (Table 2) indicate that any J-domain could associate with BiP at any given time. However, a large proportion of BiP will be engaged with polypeptide substrates, suggesting that under steady-state conditions, the specific affinities become relevant.[72]

The Type I membrane protein ERj1 can complement the lethal deletion of Sec63p in *Saccharomyces cerevisiae* even though there is no ERj1 orthologue in yeast.[74,75] ERj1 recruits BiP to translating ribosomes across the ER membrane. Moreover, in the presence of ERj1, BiP can be directly crosslinked to the nascent chain. The ribosome interaction is mediated by the large cytosolic domain of ERj1 and is regulated by BiP; in the absence of BiP, this interaction inhibits translational initiation. Based on these observations and the fact that ERj1 can be crosslinked to nascent chains, a role for ERj1 in cotranslational protein transport into the ER has been postulated (see below).[76,77] Because the cytosolic domain of ERj1 contains two Myb domains typically found in transcriptional regulators, Zupicich et al predicted ERj1 as a membrane-tethered transcription factor.[78] Indeed, the cytosolic domain of ERj1 can enter the nucleus, which is mediated by the nuclear import receptor importin ß.[76,78] Thus, ERj1 displays all the characteristics of a UPR mediator, the ability to bind to BiP and to regulate gene expression. ERj1 might modulate the ER-associated protein biogenesis at different levels, both at the level of translation and at the level of transcription.[76]

The multispanning membrane protein ERj2/Sec63 has been found in equimolar amounts as the heterotrimeric Sec61 complex.[79] Furthermore, a direct association between the ERj2/Sec63 and the Sec61 translocase has been shown, indicating a function for ERj2/Sec63 in protein transport into the ER (see below).[79,80] The two ER-resident calcium-binding proteins, calumenin and reticulocalbin, were identified as binding partners for ERj2/Sec63. These associations might be related to a role for ERj2/Sec63 in the maintenance of calcium homeostasis in the ER by controlling calcium ion efflux from the ER during protein translocation.[81] Moreover, it was shown that mutations in *SEC63* cause autosomal dominant polycystic liver diseases and small bowel cancer in humans (see chapter by Müller et al).[82]

The soluble lumenal ERj3, homologue of the yeast Scj1p, is the ERj family member most similar to the archetypical DnaJ protein. In addition to a J-domain, ERj3 contains a glycine/phenylalanine-rich region followed by a cysteine-rich domain.[68,84] It interacts directly with a number of nascent unfolded and mutant secretory proteins and recruits BiP to unfolded substrates.[83,85] Furthermore, ERj3 was found in direct association with SDF2-L1 (stromal cell-derived factor 2-like protein 1), a protein similar to O-mannosyltransferase, indicating a role for ERj3 in O-glycosylation of proteins in the ER.[86,87] Interestingly, these two proteins may exist not only as a heterodimer but also appear to be parts of a larger chaperone complex associated with unfolded IgG heavy chains, which includes BiP, Grp170, Grp94 (Hsp90 homologue of the ER), CaBP1 (calcium-binding protein 1), UDP-glucosyltransferase, ERp72 (thiol oxidoreductase), PDI (protein disulfide isomerase) and cyclophilin B (peptidyl-prolyl isomerase), in addition to ERj3 and SDF2-L1, indicating a more general role for ERj3 in the folding and maturation of newly synthesized proteins. Except for ERj3, this complex also forms in the absence of nascent protein synthesis and has been found in a variety of cell types. This suggests that instead of existing as free pools, ER chaperones are integrated in a precast ER network that can bind to unfolded protein substrates.[85]

The Type I membrane-anchored protein ERj4 is weakly expressed under nonstress conditions (Table 1), whereas it is highly induced at both the mRNA and protein levels in response to ER-stress.[69] Like ERj3, ERj4 interacts directly with folding substrates. This interaction is independent of the J-domain and most likely involves the glycine/phenylalanine-rich region, which might also be true for ERj3.[69,88] ERj4 is required for ERAD of nonglycosylated proteins. Together with misfolded substrates, it coprecipitates with the cytosolic ERAD component p97, an AAA-ATPase, indicating that ERj4 remains associated with misfolded proteins until it is retrotranslocated to the cytosol.[88] Furthermore, ERj4 prevents ER stress-induced apoptosis.[89]

In addition to the J-domain, the soluble lumenal ERj5 protein has four thioredoxin domains, characteristic of PDIs.[70,90] Therefore, ERj5 is expected to link protein folding/unfolding to disulfide bridge formation/reduction. Downregulation of *ERJ5* expression enhances ER retention and inhibits degradation of misfolded proteins, indicating a role for ERj5 in ERAD. Accordingly, a BiP-dependent direct interaction with nonglycosylated ERAD substrates has been detected.[88] It is likely that in conjunction with BiP, the J-domain of ERj5 mediates the maintenance of an unfolded conformation, while the thioredoxin domains act similarly to PDI in disulfide bond reduction prior to the retrotranslocation of misfolded proteins.[91] ERj5 may also function as a redox sensor to detect mispaired disulfide bonds and target these proteins for degradation by ERAD.[70,88,90] Like ERj4, ERj5 coprecipitates with p97, suggesting that ERj4 and ERj5 are part of a larger lumenal quality control complex that targets misfolded proteins for retrotranslocation and degradation by the proteasome.[88]

The soluble ERj6/p58^{IPK} protein contains a carboxy-terminal J-domain and nine tetratricopeptide repeats (TPRs). ERj6/p58^{IPK} was long thought to reside in the cytosol and to be a peripheral protein of the ER membrane. Recent analyses show that it contains a cleavable amino terminal signal peptide and is located in the ER lumen, where it interacts with BiP in a J-domain-dependent manner and may function as a BiP co-chaperone to optimize protein folding homeostasis in the ER.[71] Furthermore, ERj6/p58^{IPK} is believed to be involved in ERAD.[88] The role of the TPR domain remains unclear. ERj6 might recruit BiP to specific substrates via the TPRs. In the cytosol, TPR domains participate in the assembly of the Hsp70-Hsp90 multi-chaperone complexes.[92] Even though the ER homologues do not contain the canonical TPR binding motif, EEVD, it is possible that ERj6/p58^{IPK} contributes to the formation of an ER chaperone network.

In accordance with their function in folding and ERAD, the expression of *ERJ3*, *ERJ4*, *ERJ5* and *ERJ6* is upregulated by the UPR, whereas *ERJ1* and *ERJ2*, which appear to be involved in protein transport into the ER, are not induced by UPR (Fig. 3).

The Role of BiP and Its Co-Chaperones in Protein Transport

BiP/Kar2p is essential for all stages of co- and posttranslational protein transport into the ER. Transport occurs in three consecutive steps: membrane association of the precursor protein; insertion into the protein-conducting translocase, the Sec61 complex; and finally, completion of translocation.

During posttranslational transport in yeast, precursor proteins synthesized in the cytosol are recognized by a receptor complex in the ER membrane.[58,93] Release from the cytosolic binding site requires the lumenal Kar2p, ATP-hydrolysis and the interaction between Kar2p and Sec63p, defining a role for these two proteins early in the ER translocation process.[58,94] In mammals, BiP and Grp170 are involved in co- and posttranslational insertion of precursor polypeptides into the Sec61 complex.[36]

BiP/Kar2p has also been identified as a lumenal protein involved in the completion of protein translocation.[57,95] The unidirectionality of transport is mediated by direct binding of BiP/Kar2p to the translocating chain. In vitro, BiP/Kar2p can be replaced by antibodies against the transport substrate or by avidin in case of a biotinylated precursor. Thus, BiP/Kar2p acts as a Brownian ratchet, which in yeast requires the interaction of Kar2p and Sec63p.[57-59,95,96]

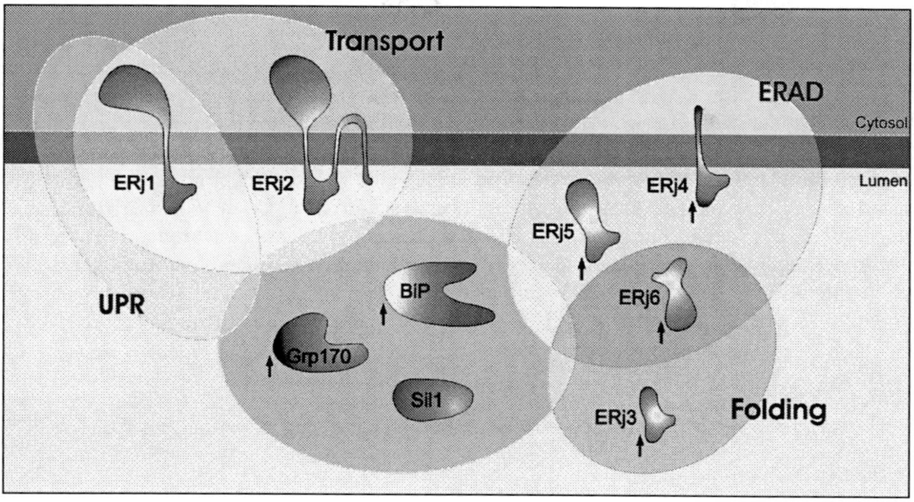

Figure 3. The Hsp70/Hsp40 network in the endoplasmic reticulum. Proteins marked by an arrow are upregulated by the unfolded protein response.

Furthermore, BiP was shown to maintain the permeability barrier between the cytosol and the ER lumen by sealing the lumenal side of the mammalian Sec61 complex in the absence of protein translocation and at several stages during cotranslational translocation of precursor proteins.[41,97] BiP was reported to cooperate with an unknown ER Hsp40 partner with a lumenal J-domain in mammalian Sec61 translocase gating; its yeast orthologue Kar2p could not substitute for BiP in this function. It is assumed that the BiP-mediated opening/closing mechanism of the translocase is similar to its mode of interaction with folding substrates. This is based on the observation that Sec61 gating requires both the substrate-binding domain and the ATPase domain of BiP, as well as unidentified membrane proteins.[97] In this context, it is worth noting that BiP is able to bind parts of the J-domain as a substrate.[98]

On the basis of analogies to yeast and the facts that human *ERJ1* can complement a deletion of the *SEC63* gene in yeast and that ERj4 appears to be absent from pancreatic rough ER under nonstress conditions, the two membrane Hsp40 co-chaperones ERj1 and ERj2/Sec63 appear to be prime candidates for alternatively cooperating with BiP in protein transport and the associated regulation of the translocase gating state. By providing a J-domain in the immediate vicinity of the Sec61 complex, both proteins can recruit BiP to the translocase and the nascent chain and simultaneously stimulate BiP's ATPase activity in order to allow a productive interaction. Furthermore, based on its capacity to combine characteristics of UPR mediators with the ability to interact with translating ribosomes and nascent chains, ERj1 could provide a direct link between the ER protein transport and the UPR.

Open Questions

Some questions about the function of BiP and its co-chaperones in protein transport remain. In yeast, there is only one Hsp40 partner of BiP known for which a clear role in transport has been demonstrated, the membrane-anchored Sec63p. In the mammalian ER, two Hsp40 membrane proteins exist, ERj1 and ERj2/Sec63, both of which seem to be engaged in translocation. Why are two proteins necessary in higher eukaryotes? Are the two proteins specialized with respect to substrate specificity, transport mode (co-versus posttranslational translocation), transport direction (import versus export), or participation in different stages of transport? Is one protein involved in the actual translocation, whereas the other mediates gating of the Sec61 translocon? Since the two proteins are abundant and not stress-induced, it can be assumed that they both function in the ER under normal conditions. However, it is also feasible that the specific roles of ERj1 and ERj2/Sec63 change under ER stress conditions, that is, when the suggested ability of ERj1 to operate in the UPR is considered. Still, it is possible that the two proteins are simply functionally redundant. One would expect at least partially overlapping functions because mutations in *SEC63* that cause hereditary autosomal dominant polycystic liver diseases are not lethal. A number of similar questions remain to be addressed for the two ER-resident nucleotide exchange factors, Grp170 and Sil1. Do they have unique roles in the stress response? While Grp170 levels are elevated under stress conditions, Sil1 is not UPR-regulated. Functional redundancy could explain the nonlethality of the neurodegenerative disease MarinescoSjögren syndrome. Answers to these questions will not only help to complete our knowledge about the dynamic process of protein transport, the involvement of Hsp70 and Hsp40 chaperones and regulatory mechanisms but will also be of great biomedical relevance.

Acknowledgements

The authors' work on this subject was supported by the Deutsche Forschungs-gemeinschaft.

References

1. Jung M, Bies C, Dudek J et al. Multiple roles of molecular chaperones in protein transport into the mammalian endoplasmic reticulum. Research Signpost, Recent Research Developments in Molecular and Cellular Biology 2003; 3:549-562.
2. Hiller MM, Finger A, Schweiger M et al. ER degradation of a misfolded luminal the cytosolic ubiquitin-proteasome pathway. Science 1996; 273:1725-1728.
3. Werner ED, Brodsky JL, McCracken AA. Proteasome-dependent endoplasmic reciculum-associated protein degradation: an unconventional route to a familiar fate. Proc Natl Acad Sci USA 1996; 93:13797-13801.
4. Buck TM, Wright CM, Brodsky JL. The activities and function of molecular chaperones in the endoplasmic reticulum. Semin Cell Dev Biol 2007; 18:751-761.
5. Ma Y, Hendershot LM. The unfolding tale of the unfolded protein response. Cell 2001; 107:827-830.
6. Zhang K, Kaufman RJ. Signaling the unfolded protein response from the endoplasmic reticulum. J Biol Chem 2004; 279:25935-25938.
7. Flaherty KM, DeLuca-Flaherty C, McKay DB. Three-dimensional structure of the ATPase fragment of a 70K heat-shock cognate protein. Nature 1990; 346:623-628.
8. Wang TF, Chang JH, Wang C. Identification of the peptide binding domain of hsc70. 18-kilodalton fragment located immediately after ATPase domain is sufficient for high affinity binding. J Biol Chem 1993; 268:26049-26051.
9. Zhu X, Zhao X, Burkholder WF et al. Structural analysis of substrate binding by the molecular chaperone DnaK. Science 1996; 272:1606-1614.
10. Hartl FU. Molecular chaperones in cellular protein folding. Nature1996; 381:571-579.
11. McCarty JS, Buchberger A, Reinstein J et al. The role of ATP in the functional cycle of the DnaK chaperone system. J Mol Biol 1995; 249:126-137.
12. Theyssen H, Schuster HP, Packschies L et al. The second step of ATP binding to DnaK induces peptide release. J Mol Biol 1996; 263:657-670.
13. Cheetham ME, Caplan AJ. Structure, function and evolution of DnaJ: conservation and adaptation of chaperone function. Cell Stress Chaperones 1998; 3:28-36.
14. Gassler CS, Buchberger A, Laufen T et al. Mutations in the DnaK chaperone affecting interaction with the DnaJ cochaperone. Proc Natl Acad Sci USA 1998; 95:15229-15234.
15. Suh WC, Burkholder WF, Lu CZ et al. Interaction of the Hsp70 molecular chaperone, DnaK, with its cochaperone DnaJ. Proc Natl Acad Sci USA 1998; 95:15223-15228.
16. Liberek K, Marszalek J, Ang D et al. Escherichia coli DnaJ and GrpE heat shock proteins jointly stimulate ATPase activity of DnaK. Proc Natl Acad Sci USA 1991; 88:2874-2878.
17. Silver PA, Way JC. Eukaryotic DnaJ homologues and the specificity of Hsp70 activity. Cell 1993; 74:5-6.
18. Kassenbrock CK, Kelly RB. Interaction of heavy chain binding protein (BiP/GRP78) with adenine nucleotides. EMBO J 1989; 81:461-1467.
19. Wall D, Zylicz M, Georgopoulos C. The N-terminal 108 amino acids of the Escherichia coli DnaJ protein stimulate ATPase activity of DnaK and are sufficient for replication. J Biol Chem 1994; 269:5446-5451.

20. Szyperski T, Pellecchia M, Wall D et al. NMR structure determination of the Escherichia coli DnaJ molecular chaperone: secondary structure and backbone fold of the N-terminal region (residues 2-108) containing the highly conserved J domain. Proc Natl Acad Sci USA 1994; 91:11343-11347.
21. Karzai AW, McMacken R. A bipartite signaling mechanism involved in DnaJ-mediated activation of the Escherichia coli DnaK protein. J Biol Chem 1996; 271:11236-11246.
22. Laufen T, Mayer MP, Beisel C et al. Mechanism of regulation of hsp70 chaperones by DnaJ cochaperones. Proc Natl Acad Sci USA 1999; 96:5452-5457.
23. Johnson JL, Craig EA. An essential role for the substrate-binding region of Hsp40s in saccharomyces cerevisiae. J Cell Biol 2001; 152:851-856.
24. Cheetham ME, Caplan AJ. Structure, function and evolution of DnaJ: conservation and adaptation of chaperone function. Cell Stress Chaperones 1998; 3:28-36.
25. Bukau B, Horwich AL. The Hsp70 and Hsp60 chaperone machines. Cell 1998; 92:351-366.
26. Höhfeld J, Jentsch S. GrpE-like regulation of the hsc70 chaperone by the anti-apoptotic protein BAG-1. EMBO J 1997; 16:6209-6216.
27. Kabani M, Beckerich JM, Gaillardin C. Sls1p stimulates Sec63p-mediated activation of Kar2p in a conformation-dependent manner in the yeast endoplasmic reticulum. Mol Cell Biol 2000; 18:6923-6934.
28. Tyson JR, Stirling CJ. LHS1 and SIL1 provide a lumenal function that is essential for protein translocation into the endoplasmic reticulum. EMBO J 2000; 19:6440-6452.
29. Chung KT, Shen Y, Hendershot LM. BAP, a mammalian BiP associated protein, is a nucleotide exchange factor that regulates the ATPase activity of BiP. J Biol Chem 2002; 277:47557-47563.
30. Shomura Y, Dragovic Z, Chang HC et al. Regulation of Hsp70 function by HspBP1: structural analysis reveals an alternate mechanism for Hsp70 nucleotide exchange. Mol Cell 2005; 17:367-379.
31. Weitzmann A, Volkmer J, Zimmermann R. The nucleotide exchange factor activity of Grp170 may explain the nonlethal phenotype of loss of Sil1 function in man and mouse. FEBS Lett 2006; 580:5237-5240.
32. Mayer M, Reinstein J, Buchner J. Modulation of the ATPase cycle of BiP by peptides and proteins. J Mol Biol 2003; 330:137-144.
33. Haas IG, Wabl M. Immunoglobulin heavy chain binding protein. Nature 1983; 306:387-389.
34. Rose MD, Misra LM, Vogel JP. KAR2, a Karyogamy Gene, is the yeast homolog of the mammalian BiP/GRP78 gene. Cell 1989; 57:1211-1221.
35. Craven RA, Egerton M, Stirling CJ. A novel Hsp70 of the yeast ER is required for the efficient translocation of a number of protein precursors. EMBO J 1996; 15:2640-2650.
36. Dierks T, Volkmer J, Schlenstedt G et al. A microsomal ATP-binding protein involved in efficient protein transport into the mammalian endoplasmic reticulum. EMBO J 1996; 15:6931-6942.
37. Normington K, Kohno K, Kozutsumi Y et al. S. cereviseae encodes an essential protein homologous in sequence and function to mammalian BiP. Cell 1989; 57:1223-1236.
38. Luo S, Mao C, Lee B et al. GRP78/BiP is required for cell proliferation and protecting the inner cell mass from apoptosis during early mouse embryonic development. Mol Cell Biol 2006; 15:5688-5697.
39. Mori K, Sant A, Kohnol K et al. A 22 bp cis-acting element is necessary and sufficient for the induction of the yeast KAR2 (BiP) gene by unfolded proteins. EMBO J 1992; 7:2583-2593.
40. Kaufman RJ. Stress signaling from the lumen of the endoplasmic reticulum: coordination of gene transcriptional and translational controls. Genes Dev 1999; 13:1211-1233.
41. Hamman BD, Hendershot LM, Johnson AE. BiP maintains the permeability barrier of the ER membrane by sealing the lumenal end of the translocation pore before and early in translocation. Cell 1998; 92:747-758.
42. Knittler MR, Haas lG. Interaction of BiP with newly synthesized immunoglobulin light chain molecules: cycles of sequential binding and release. EMBO J 1992; 4:1573-1581.
43. Molinari M, Galli C, Piccaluga V et al. Sequential assistance of molecular chaperones and transient formation of covalent complexes during protein degradation from the ER. J Cell Biol 2002; 158:247-257.
44. Bertolotti A, Zhang Y, Hendershot LM et al. Dynamic interaction of BiP and ER stress transducers in the unfolded-protein response. Nat Cell Biol 2000; 2:326-332.
45. Lievremont JP, Rizzuto R, Hendershot L et al. BiP, a major chaperone protein of the endoplasmic reticulum lumen, plays a direct and important role in the storage of the rapidly exchanging pool of Ca^{2+}. J Biol Chem 1997; 272:30873-30879.
46. Chen X, Easton D, Oh HJ et al. The 170 kDa glucose regulated stress protein is a large HSP70-, HSP110-like protein of the endoplasmic reticulum. FEBS Lett 1996; 380:68-72.
47. Easton DP, Kaneko Y, Subjeck JR. The Hsp110 and Grp170 stress proteins: newly recognized relatives of the Hsp70s. Cell Stress Chaperones 2000; 5:276-290.
48. Shaner L, Morano KA. All in the family: atypical Hsp70 chaperones are conserved modulators of Hsp70 activity. Cell Stress Chaperones 2007; 12:1-8.

49. Weitzmann A, Baldes C, Dudek J et al. The heat shock protein 70 molecular chaperone network in the pancreatic endoplasmic reticulum—a quantitative approach. FEBS J 2007; 274:5175-87.
50. Rothblatt JA, Deshaies RJ, Sanders SL et al. Multiple genes are required for proper insertion of secretory proteins into the endoplasmic reticulum in yeast. J Cell Biol 1989; 109:2641-2652.
51. Eki T, Naitou M, Hagiwara H et al. Analysis of a 36.2 kb DNA sequence including the right telomere of chromosome VI from Saccharomyces cerevisiae. Yeast 1996; 12:146-167.
52. Famá MC, Raden D, Zacchi N et al. The saccharomyces cerevisiae YFR041C/ERJ5 gene encoding a type I membrane protein with a J domain is required to preserve the folding capacity of the endoplasmic reticulum. Biochim Biophys Acta 2007; 1773:232-242.
53. Schlenstedt G, Harris S, Risse B et al. A yeast DnaJ homologue, Scj1p, can function in the endoplasmic reticulum with BiP/Kar2p via a conserved domain that specifies interactions with Hsp70s. J Cell Biol 1995; 129:979-988.
54. Nishikawa S, Endo T. The yeast jem1p is a DnaJ-like protein of the endoplasmic reticulum membrane required for nuclear fusion. J Biol Chem 1997; 272:12889-12892.
55. Deshaies RJ, Sanders SL, Feldheim DA et al. Assembly of yeast Sec proteins involved in translocation into the endoplasmic reticulum into a membrane-bound multisubunit complex. Nature 1991; 349:806-808.
56. Brodsky JL, Schekman R. A Sec63-BiP complex is required for protein translocation in a reconstituted proteoliposome. J Cell Biol 1993; 123:1355-1263.
57. Lyman SK, Schekman R. Interaction between BiP and Sec63p is required for the completion of protein translocation into the ER of saccharomyces cerevisiae. J Cell Biol 1995; 131:1163-1171.
58. Lyman SK, Schekman R. Binding of secretory precursor polypeptides to a translocon subcomplex is regulated by BiP. Cell 1997; 88:85-96.
59. Corsi AK, Schekman R. The lumenal domain of Sec63p stimulates the ATPase activity of BiP and mediates BiP recruitment to the translocon in saccharomyces cerevisiae. J Cell Biol 1997; 137:1483-1493.
60. Brodsky JL, Goeckeler J, Schekman R. BiP and Sec63p are required for both co- and posttranslational protein translocation into the endoplasmic reticulum. Proc Natl Acad Sci USA 1995; 92:9643-9646.
61. Young BP, Craven RA, Reid PJ et al. Sec63p and Kar2p are required for the translocation of SRP-dependent precursors into the yeast endoplasmic reticulum in vivo. EMBO J 2001; 20:262-271.
62. Plemper RK, Bömler S, Bordallo J et al. Mutant analysis links the translocon and BiP to retrograde protein transport for ER degradation. Nature 1997; 388:891-895.
63. Brodsky JL, Werner ED, Dubas ME et al. The requirement for molecular chaperones during endoplasmic reticulum-associated protein degradation demonstrates that protein export and import are mechanistically distinct. J Biol Chem 1999; 274:3453-3460.
64. Silberstein S, Schlenstedt G, Silver PA et al. A role for the DnaJ homologue Scj1p in protein folding in the yeast endoplasmic reticulum. J Cell Biol 1998; 143:921-933.
65. Nishikawa S, Fewell SW, Kato Y et al. Molecular chaperones in the yeast endoplasmic reticulum maintain the solubility of proteins for retrotranslocation and degradation. J Cell Biol 2001; 153:1061-1070.
66. Brightman SE, Blatch GL, Zetter BR. Isolation of a mouse cDNA encoding MTJ1, a new murine member of the DnaJ family of proteins. Gene 1995; 153:249–254.
67. Skowronek MH, Rotter M, Haas IG. Molecular characterization of a novel mammalian DnaJ-like Sec63p homolog. Biol Chem 1999; 380:1133-1138.
68. Bies C, Guth S, Janoschek K et al. A Scj1p homolog and folding catalysts present in dog pancreas microsomes. Biol Chem 1999; 380:1175-1182.
69. Shen Y, Meunier L, Hendershot LM. Identification and characterization of a novel endoplasmic reticulum (ER) DnaJ homologue, which stimulates ATPase activity of BiP in vitro and is induced by ER stress. J Biol Chem 2002; 277:15947-15956.
70. Cunnea PM, Miranda-Vizuete A, Bertoli G et al. ERdj5, an endoplasmic reticulum (ER)-resident protein containing DnaJ and thioredoxin domains, is expressed in secretory cells or following stress. J Biol Chem 2003; 278:1059-1066.
71. Rutkowski DT, Kang SW, Goodman AG et al. The role of p58IPK in protecting the stressed endoplasmic reticulum. Mol Biol Cell 2007; 18:3681-3691.
72. Weitzmann A, Baldes C, Dudek J et al. The heat shock protein 70 molecular chaperone network in the pancreatic endoplasmic reticulum—a quantitative approach. FEBS J 2007; 274:5175-5187.
73. Guth S, Völzing C, Müller A et al. Protein transport into canine pancreatic microsomes: a quantitative approach. Eur J Biochem 2004; 271:3200-3207.
74. Kroczynska B, Evangelista CM, Samant SS et al. The SANT2 domain of murine tumor cell DnaJ-like protein 1 human homologue interacts with α1-antichymotrypsin and kinetically interferes with its serpin inhibitory activity. J Biol Chem 2004; 279:11432-11443.
75. Dudek J, Volkmer J, Bies C et al. A novel type of cochaperone mediates transmembrane recruitment of DnaK-like chaperones to ribosomes. EMBO J 2002; 21:2958-2967.

76. Dudek J, Greiner M, Müller A et al. ERj1p plays a basic role in protein biogenesis at the endoplasmic reticulum. Nat Struct Mol Biol 2005; 12:1008-1014.
77. Blau M, Mullapudi S, Becker T et al. ERj1p uses a universal ribosomal adaptor site to coordinate the 80S ribosome at the membrane. Nat Struct Mol Biol 2005; 12:1015-1016.
78. Zupicich J, Brenner SE, Skarnes WC. Computational prediction of membrane-tethered transcription factors. Genome Biol 2001; 2:50.1-50.6.
79. Tyedmers J, Lerner M, Bies C et al. Homologs of the yeast sec complex subunits Sec62p and Sec63p are abundant proteins in dog pancreas microsomes. Proc Natl Acad Sci USA 2000; 97:7214-7219.
80. Meyer HA, Grau H, Kraft R et al. Mammalian sec61 is associated with sec62 and sec63. J Biol Chem 2000; 275:14550-14557.
81. Tyedmers J, Lerner M, Nastainczyk W et al. Calumenin and reticulocalbin are associated with the protein translocase of the mammalian endoplasmic reticulum. J Biol Sci 2005; 5:70-75.
82. Davila S, Furu L, Gharavi AG et al. Mutations in SEC63 cause autosomal dominant polycystic liver disease. Nat Genet 2004; 36:575–577.
83. Shen Y, Hendershot LM. ERdj3p, a stress-inducible endoplasmic reticulum DnaJ homologue, serves as a cofactor for BiP's interactions with unfolded substrates. Mol Biol Cell 2004; 16:40-50.
84. Nicoll WS, Botha M, McNamara C et al. Cytosolic and ER J-domains of mammalian and parasitic origin can functionally interact with DnaK. Int J Biochem Cell Biol 2007; 39:736-751.
85. Meunier L, Usherwood YK, Chung KT et al. A subset of chaperones and folding enzymes form multiprotein complexes in endoplasmic reticulum to bind nascent proteins. Mol Biol Cell 2002; 13:4456-4469.
86. Bies C, Blum R, Dudek J et al. Characterization of pancreatic ERj3p, a homolog of yeast DnaJ-like protein Scj1p. Biol Chem 2004; 385:389-395.
87. Fukuda S, Sumii M, Masuda Y et al. Murine and human SDF2L1 is an endoplasmic reticulum stress-inducible gene that encodes a new member of the pmt/rt protein family. Biochem Biophys Res Commun 2001; 280:407-414.
88. Dong M, Bridges JP, Apsley K et al. ERdj4 and ERdj5 are required for endoplasmic ssociated protein degradation of misfolded surfactant protein C. Mol Biol Cell 2008; 19:2620-2630.
89. Kurisu J, Honma A, Miyajima H et al. MDG1/ERdj4, an ER-resident DnaJ family member, suppresses cell death induced by ER stress. Genes Cells 2003; 8:189-202.
90. Hosoda A, Kimata Y, Tsuru A et al. JPDI, a novel endoplasmic reticulum-resident protein containing both a BiP-interacting J-domain and thioredoxin-like motifs. J Biol Chem 2003; 278:2669-2676.
91. Tsai B, Rodighiero C, Lencer WI et al. Protein disulfide isomerise acts as a redox-dependent chaperone to unfold cholera toxin. Cell 2001; 104:937-948.
92. Brinker A, Scheufler C, Von Der Mulbe F et al. Ligand discrimination by TPR domains. Relevance and selectivity of EEVD-recognition in Hsp70 × Hop × Hsp90 complexes. J Biol Chem 2002; 277:19265-19275.
93. Müsch A, Wiedmann M, Rapoport TA. Yeast Sec proteins interact with polypeptides traversing the endoplasmic reticulum membrane. Cell 1992; 69:343-352.
94. Matlack KE, Plath K, Misselwitz B et al. Protein transport by purified yeast sec complex and kar2p without membranes. Science 1997; 277:938-941.
95. Tyedmers J, Lerner M, Wiedmann M et al. Polypeptide chain binding proteins mediate completion of cotranslational protein translocation into the mammalian endoplasmic reticulum. EMBO Rep 2003; 4:505-510.
96. Matlack KE, Misselwitz B, Plath K et al. BiP acts as a molecular ratchet during posttranslational transport of prepro-α-factor across the ER membrane. Cell 1999; 97:553-564.
97. Alder NN, Shen Y, Brodsky JL et al. The molecular mechanism underlying BiP-mediated gating of the sec61 translocon of the endoplasmic reticulum. J Cell Biol 2005; 168:389-399.
98. Hennessy F, Nicoll WS, Zimmermann R et al. Not all J domains are created equal: implications for the specificity of Hsp40-Hsp70 interactions. Protein Sci 2005; 14:1697-709.

Chapter 7

Membrane Protein Biosynthesis at the Endoplasmic Reticulum

Benedict C.S. Cross* and Stephen High*

Abstract

Current estimates suggest that 20-30% of a typical eukaryotic genome encodes proteins that span a membrane, the majority of which are generated at the endoplasmic reticulum (ER). A remarkable proteinaceous complex embedded in the membrane of the ER, termed the ER translocon, is responsible for both the transport of water-soluble proteins from the cytosol into the ER lumen and for the incorporation of membrane proteins into the lipid bilayer. Once synthesised, proteins may be trafficked from the ER to their final subcellular locations or secreted from the cell. Thus, for many soluble proteins, the ER marks the entry point into the secretory pathway and the site at which significant and essential modifications to the structure of the protein are carried out. Whilst some of the basic principals underlying the biosynthesis of integral membrane proteins are similar to those for soluble proteins, membrane protein biosynthesis is complicated by the presence of one or more membrane-spanning regions. This review will focus on our current understanding of translocon-mediated membrane protein integration at the mammalian ER, with a particular emphasis on the biogenesis of complex polytopic proteins with multiple transmembrane (TM) segments.

Introduction

Proteins destined for the ER carry distinct motifs or signals within the polypeptide chain that direct them to the membrane of the ER. A typical N-terminal signal sequence is between 15 and 30 residues in length and has a tripartite organisation comprising a core of hydrophobic residues flanked by n- and c-regions of more polar amino acids (see chapter by Kapp et al for a detailed discussion of signal sequences). N-terminal signal sequences that play a purely targeting role are usually removed as part of the maturation process during protein translocation across the ER membrane in a process catalysed by the signal peptidase complex (SPC) in the lumen of the ER. Signal sequences with a cleavage consensus motif in the c-region are recognised by SPC, which removes the signal sequence and releases it into the lipid bilayer (Fig. 1A). Membrane proteins with a cleavable signal sequence are anchored in the membrane by virtue of a stop-transfer sequence, a second stretch of hydrophobic residues that is prevented from entering the ER lumen and rather is incorporated into the lipid bilayer (see Fig. 1B). Alternatively, membrane proteins without cleavable signal sequences are targeted to the ER by a dual-function signal anchor sequence, which both identifies the protein as ER-destined and forms a TM segment within the membrane (see Fig. 1C,D).

The first ER targeting sequence also defines the orientation of proteins within the membrane as charged residues directly adjacent to the hydrophobic core of the targeting sequence are typically retained in the cytosol by a phenomenon termed the "positive-inside rule" (see Fig. 1).[1] Hence,

*Corresponding Authors: Benedict C.S. Cross and Stephen High—Faculty of Life Sciences, University of Manchester, Manchester, M13 9PT, UK.
Emails: benedict.cross@manchester.ac.uk and stephen.high@manchester.ac.uk

Protein Transport into the Endoplasmic Reticulum, edited by Richard Zimmermann

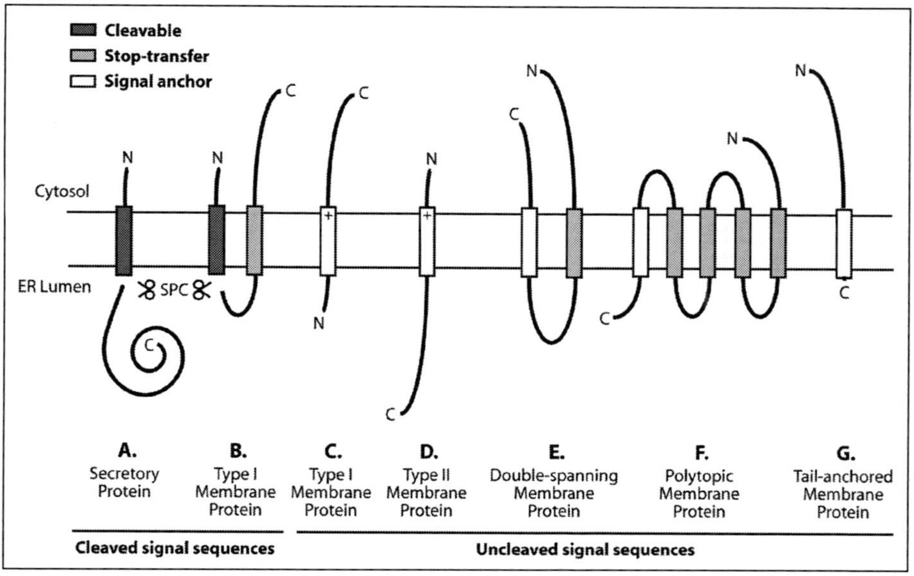

Figure 1. Membrane protein topology. A) Classical secretory protein. Cleavable signal sequences (dark grey) are removed in the lumen of the ER are by the signal peptide peptidase (SPC) complex. The protein is then released into the lumen of the ER for additional processing and trafficking. B) Single-spanning integral membrane proteins with cleavable signal sequence have a subsequent stop-transfer sequence (light grey) and thus result in a Type I membrane topology. C) Integral membrane proteins with an uncleaved signal sequence, i.e., a signal-anchor sequence (white), are of Type I orientation when the charged residues are C-terminal of the signal-anchor sequence. D) A Type II membrane orientation is adopted if charged residues are present N-terminal to the signal-anchor sequence. E) Double spanning membrane proteins will have both their N- and C-termini on the same side of the membrane. F) Polytopic membrane proteins with uncleaved signal sequences may have combinations of both signal-anchor and stop-transfer sequences. G) Tail-anchored proteins have a Type II orientation resulting from a C-terminal targeting sequence.

if a preponderance of basic residues is present at the C-terminus of the signal sequence, a Type I transmembrane orientation is favoured (where the N-terminus is in the lumen; see Fig. 1C), whilst a higher proportion of charged residues at the N-terminus of the signal sequence results in a Type II orientation (N-terminus in the cytosol; see Fig. 1D). Single-spanning membrane proteins with a cleavable ER targeting signal will invariably adopt a Type I orientation, owing to the necessity to translocate the SPC cleavage motif into the ER lumen. In the case of polytopic membrane proteins, the orientation of the first TM segment will typically define subsequent TM segment orientations, such that an alternating arrangement is achieved (see Fig. 1F).

A complex termed the signal recognition particle (SRP) is responsible for the delivery of most proteins that are destined for the ER. Mammalian SRP consists of six protein subunits (SRP9, SRP14, SRP19, SRP, 54, SRP68 and SRP72) bound to a core ~300 nucleotide RNA backbone[2].[3] SRP is remarkably well conserved across kingdoms and the minimal components of the SRP54 subunit, together with at least a portion of the RNA molecule, can be found in all domains of life.[4]

Following the initiation of translation on the ribosome and the emergence of the nascent protein from the exit tunnel, SRP recognises and binds the hydrophobic signal sequence in a flexible, methionine-rich binding grove in the SRP54 subunit (see Fig. 2).[5,6] A temporary attenuation of translation is then triggered by SRP[7,8] and the whole complex of the ribosome, SRP and the nascent polypeptide chain is recruited to the ER membrane by the interaction of SRP with the

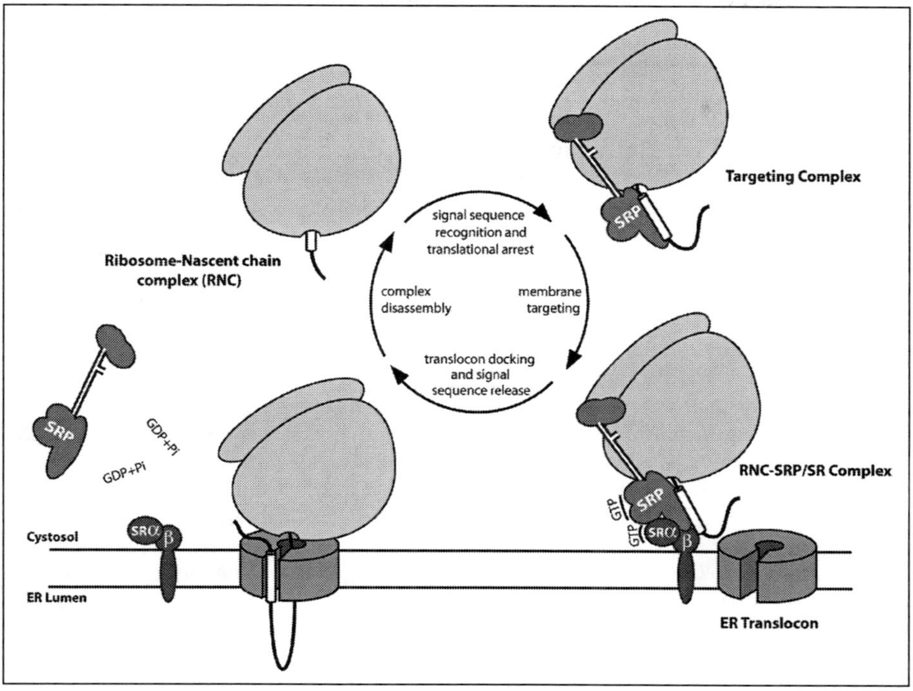

Figure 2. The SRP-dependent targeting cycle. Proteins destined for the ER are recognised and bound by the signal recognition particle (SRP) by virtue of a hydrophobic signal sequence (white) present in the protein sequence. SRP elicits a temporary attenuation in translation and the whole complex of the SRP, ribosome and nascent protein chain are recruited to the ER membrane by the interaction of SRP with SRP receptor (SR). SRP and SRα must be in the GTP bound state for this interaction to occur and GTPase domains in both proteins facilitate reciprocal GTP hydrolysis. Structural rearrangements that accompany GTP hydrolysis allow the transfer of the ribosome-nascent chain complex from SRP-SR to the ER translocon and cause the disassembly of the SRP-SR complex. SRP is then free to facilitate further rounds of targeting whilst translation resumes at the ribosome and drives translocation of the protein into the ER lumen.

membrane-bound SRP receptor (SR; see Fig. 2). This transient slowing of translation compensates for the association of the ribosome-SRP complex with a limiting population of SR molecules on the ER membrane[8] and thus prevents the synthesis of complex folding or aggregation-prone domains of the nascent protein that would be difficult to translocate across the ER membrane. This targeting pathway is therefore strictly cotranslational and is distinct from the primarily post-translational targeting of proteins to organelles such as mitochondria, nuclei, peroxisomes and chloroplasts. A special class of C-terminally anchored membrane proteins (see Fig. 1G; see also chapter by Borgese et al) are targeted posttranslationally to the ER[9] and some secretory proteins can clearly be posttranslationally targeted in yeast.[10] Whilst the prevalence of SRP-independent pathways in mammalian cells is presently unclear, current dogma suggests that most proteins are targeted cotranslationally to the mammalian ER[4] via the SRP-dependent route.

Mammalian SR comprises a soluble SRα subunit in complex with an ER-resident integral membrane protein, SRβ. SRP and SR interact with a high affinity when both are in the GTP-bound state and this close association stimulates an elegant reciprocal GTP hydrolysis reaction between the SRP54 and the SRα subunits. GTP hydrolysis promotes the disassembly of the SRP-SR complex and recycles the components for further rounds of targeting.[11] The conformational rearrangements in the SRP-SR complex that lead to hydrolysis then promote the transfer of the ribosome/cargo protein to the ER translocon where translocation of the protein into the ER lumen and TM segment integration is coordinated (Fig. 2).[12]

The Membrane Protein Integration Molecular Machinery

The ER translocon forms an aqueous protein-conducting channel through the ER membrane that catalyses both the translocation of protein into the ER lumen and the integration of proteins into the membrane. The core mammalian translocon is made up of the heterotrimeric Sec61 complex (Sec61α, Sec61β and Sec61γ).[13] In *Saccharomyces cerevisiae*, the α, β and γ subunits are termed Sec61p, Sbh1p and Sss1p respectively whilst in bacteria the equivalent subunits are SecY, SecG and SecE respectively. Importantly, the translocon complexes in these different species share both sequence identity and functional roles.[13] High-resolution structural analysis of the SecYEβ heterotrimer of the archaebacteria *Methanococcus jannaschii* has been treated as a paradigm for understanding how the Sec61 complex may form a channel through the lipid bilayer.[14] In this structure, the α subunit, which is the largest of the three and contains ten TM helices, is arranged such that a roughly hour glass-shaped pore may provide an aqueous environment which traverses the membrane (Fig. 3A,C; see also chapter Wagner et al, respectively for detailed discussions of translocon structure).[14] The pore is partially occluded on the luminal side of the membrane by a short sub-domain of TM2 in the α subunit (Fig. 3), a feature that presents a potential barrier to translocation. To overcome this, the authors invoke a plug displacement model in which this short helix may be repositioned during translocation to allow the passage of the polypeptide (see Fig. 3C).[14] This hypothesis is supported by biochemical analyses of Sec61 complexes[15-17] but is somewhat at odds with strong evidence that the gating of the translocon channel is fulfilled by an ER luminal chaperone, BiP.[18,19] The apparently important maintenance of specific ionic conditions within the ER[20] is potentially jeopardised by the presence of a large, aqueous channel through the membrane and understanding the precise contribution of the intrinsic translocon plug and BiP to translocon gating will require further studies.

X-ray structural analysis also reveals a central constriction of diameter ~5-8Å within the SecYEG pore complex that is lined with highly conserved, hydrophobic residues.[14] The authors suggest this constriction may help to seal the inactive translocon[17] and form a gasket around the translocating polypeptide, a hypothesis supported by cross-linking of a translocating polypeptide to these residues.[21] However, the dimensions of the pore derived from this structure are only wide enough to accommodate an unfolded polypeptide chain or a single alpha helix of a TM segment and is considerably narrower than previous estimates of a ~40Å diameter pore calculated using fluorescence quenching experiments.[22] It is in fact likely that the active translocon exists as an oligomer of the core Sec61 heterotrimer (see Fig. 4), a view supported by both structural studies[23,24] and biochemical data.[25,26] It may be that during cotranslational translocation, additional Sec61 heterotrimers help to stabilise the ribosome/translocon interaction in a similar manner to that suggested for the association of the bacterial SecYEG complex and the posttranslational SecA machinery.[27] Oligomerisation of the Sec61 heterotrimers may also provide the basis for a larger, consolidated pore that may account for the apparent conflict between pore diameter measurements (see Fig. 4B,C).[28,29] However, the precise stoichiometry and arrangement of the Sec61 complex has not yet reached consensus.

A further role for the higher order assembly of the Sec61 complex may be to recruit additional ER components closely associated with the translocation process. Several large enzymatic complexes associate with the ER translocon including the SPC and the oligosaccaryl transferase (OST) complex, which is responsible for the addition of carbohydrate chains to polypeptide chains with an appropriate consensus sequence. N-linked glycosylation and signal sequence cleavage can both occur cotranslocationally, hence, although not pivotal to the translocation reaction per se, a close association with the translocon seems essential. The translocating-chain associated membrane (TRAM) protein is found adjacent to several nascent polypeptides[30-32] and has been shown in some cases to stimulate translocation.[33-35] In addition, TRAM may assist the

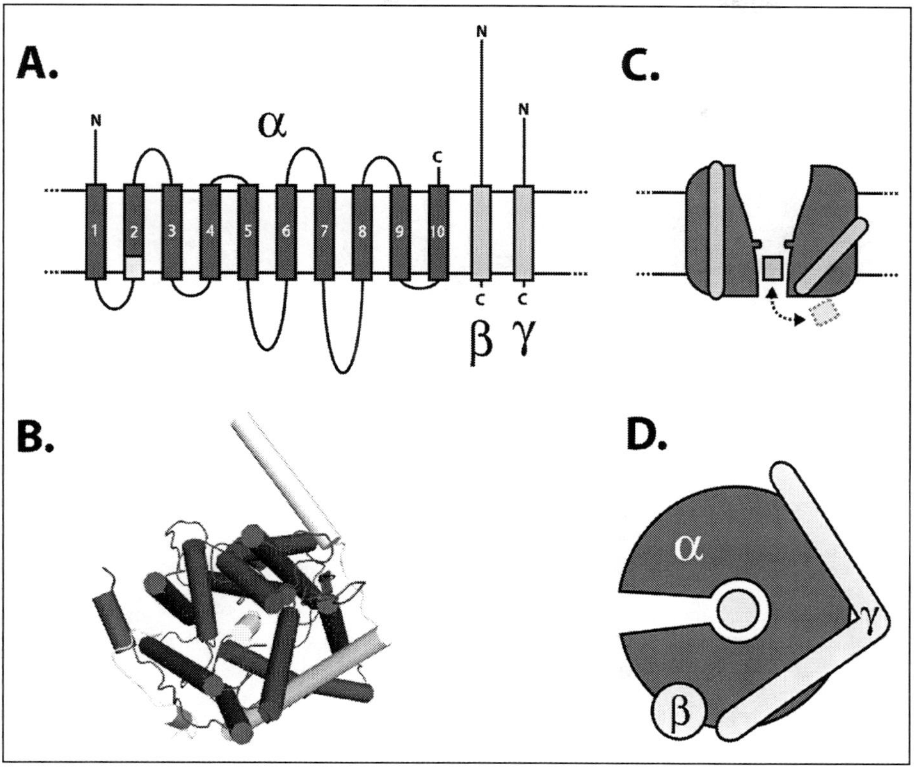

Figure 3. Translocon structure. A) Topology and domain arrangement of the mammalian Sec61 complex. A sub-domain within TM2 of Sec61α corresponds to the putative plug region (shown in white) observed in a high-resolution structure of a related bacterial translocon (see ref. 14). B) X-ray crystal structure of *M. jannaschii* SecYEβ complex (coordinates from ref. 14) viewed from the cytosol. The SecY subunit (homologous to Sec61α in mammals) is arranged with an aperture at the centre blocked by the plug region of TM2 (white). TM1-TM5 and TM6-TM10 form two halves of the complex hinged by SecE (homologous to Sec61γ in mammals). C) Cartoon view of a section the translocon showing the approximately hour-glass-shaped channel formed by a single Sec61 complex, with pore-ring constriction and proposed plug (white). This plug may be displaced during translocation to allow the passage of the polypeptide chain. D) Cartoon view of the translocon viewed from the cytosol. The two halves of the α-subunit may open to form a lateral gate through which a TM segment or signal sequence may enter the lipid phase of the membrane.

integration of membrane proteins into the lipid bilayer, perhaps even providing a chaperone-like function for TM segments.[31] Similarly, the TRAP complex, PAT-10, the Sec62/63 complex and numerous luminal chaperones may all transiently interact with the Sec61 complex depending on the specific status of a substrate and its requirements for processing at the translocon.[32,36-38] The ER translocon may also provide a central locus at which a high concentration of both ER luminal and cytosolic chaperones is maintained, thereby helping to promote correct protein folding during synthesis. Some studies also suggest that the Sec61 complex may play a central role in the removal of misfolded proteins from the ER in a process of quality control (for a review see ref. 39). The ER translocon therefore coordinates multiple crucial processes within the ER and is an incredibly dynamic structure, the core of which is clearly the Sec61 complex, but which may comprise multiple distinct combinations of components depending on the specific biosynthetic processes

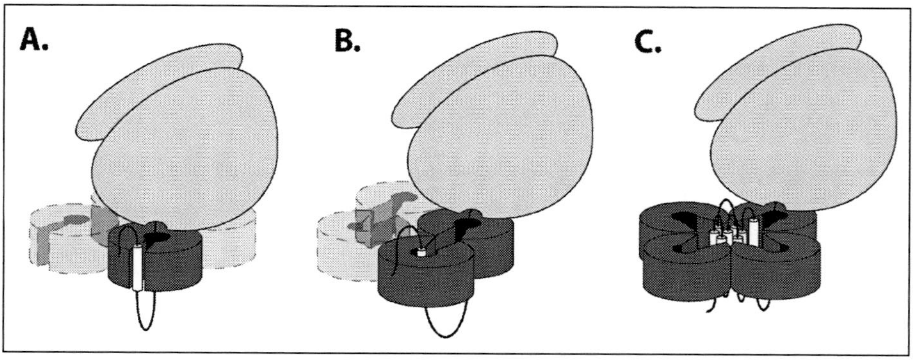

Figure 4. Membrane protein integration models. A) A single copy of the Sec61 heterotrimer forms the protein-conducting channel through the membrane. The lateral exit gate in the complex faces directly into the membrane whilst additional copies of the heterotrimer may help to stabilise the ribosome-translocon interaction or recruit additional translocon-associated components (the back-to-back model; see refs. 14 and 23). B) Two Sec61 heterotrimers are arranged face-to-face, with the lateral gates forming a consolidated large pore. The TM segments may be accommodated by one heterotrimer, whilst translocation of the soluble portion into the ER lumen may be through a second heterotrimer (see refs. 28 and 29). C) Four copies of the Sec61 heterotrimer are predicted to be present in the active translocon. In this model they are arranged such that the lateral gates face toward a central, lipid-filled cavity into which membrane proteins may be released (see refs. 24, 52, 69 and 70).

ongoing at the ER. In particular, the accessory components recruited to the ER translocon when a membrane protein is presented to the Sec61 complex might well be distinct from those required for secretory protein biogenesis.

TM Segment Partitioning into the Lipid Bilayer

In addition to forming a channel through the ER membrane from the cytosol to the lumen, the ER translocon must also have a mechanism for allowing the incorporation of TM segments into the hydrophobic environment of the lipid bilayer. Hence, when a TM segment arrives at the translocon during cotranslational translocation it must be recognised and its translocation into the lumen prevented. In order to achieve this, the ER translocon operates in a second plane and is able to open laterally with respect to the lipid bilayer (Figs. 3 and 4). Clear evidence for this model was provided using an in vitro approach in which the biosynthesis of model proteins was carefully analysed using stalled translation/translocation intermediates to simulate different stages of protein biosynthesis. Photoactivatable cross-linking reagents, incorporated into the nascent protein chain itself, can then be used to monitor the environment of the nascent polypeptide chain.[39] Using this technique it was found that whilst a secretory protein was primarily adjacent to the proteinaceous components of the ER translocon,[40] the TM segment of a membrane protein was more exposed to a lipid environment.[39] On the other hand, the soluble region of the same model membrane protein remained primarily adjacent to the protein components of the translocon,[39] presumably within the aqueous pore through the membrane.[41] High-resolution structural analysis of the archaeal translocon suggests a mechanism by which such lipid access may occur.[14] The core heterotrimer appears to be arranged such that the helices of the α-subunit form two distinct domains that are clamped together by the contiguous γ-subunit (Fig. 3B,D). It is suggested therefore, that in order to allow TM segment partitioning into the membrane the two halves of the α-subunit may open, with the γ-subunit effectively acting as a hinge, exposing the TM segment to the lipid environment (Fig. 4). This model is supported by cross-linking analyses that mapped the precise positioning of the hydrophobic segment of a translocating protein within the Sec61 complex.[42-44] TM2 and TM7 of the α-subunit in the Sec61 complex were found to be directly adjacent to the nascent protein[43] and this region in the structural model of the translocon represents the mouth of the lateral gate in the translocon. In addition, mutations in these regions of the Sec61 complex can cause perturbations in membrane topology or translocation defects,[45] supporting the importance of this region in lateral gating.

TM segment exit from the translocon pore is therefore most likely via this gate in the translocon structure, but how is the lateral release controlled? Since the signal sequence of a secretory protein displays less efficient crosslinking to membrane lipid than a generally more hydrophobic TM segment, it has been suggested that simple net hydrophobicity of the region of polypeptide may determine its propensity for membrane integration.[39] In support of this idea, the introduction of charged residues into a well defined TM segment results in a reduced efficiency of partitioning into the membrane.[31] Hence, a polypeptide span might enter the lipid phase only if this is a thermodynamically favourable event. This concept of "passive partitioning" of TM segments into the membrane was further examined by empirically defining a biological hydrophobicity scale for membrane integration, using a systematic in vitro analysis.[46] In this study, a panel of TM segments, artificially introduced into a well-characterised membrane protein, were used to determine the specific contribution of each amino acid to the integration process. The scale derived from the analyses represents an index that may quantitatively describe the tendency for membrane insertion according to the specific sequence of the TM segment. Hence, if a TM segment contains a large number of charged or polar residues, incorporation into the membrane is unfavourable. Although the relative rank contribution that each amino acid makes to the overall propensity for integration that were calculated in this study are similar to previous biochemical estimates,[46,47] a clear positional dependence was observed for the effect of specific amino acids within the TM segment.[46] For instance, a proline may be accommodated more easily at the N-terminus of the synthetic TM segment than at the C-terminus and highly charged and polar residues are more easily accommodated in the centre of the segment.[46,48] This approach clearly demonstrates the importance of the properties of the TM segment during integration and suggests that lipid-protein interactions are a major determinant of the propensity for membrane integration. These data also strongly support the idea that the translocon is laterally open to the lipid bilayer, at least transiently.[14,39] However, whilst the positional dependence observed for certain residues during integration may reflect the physical properties of the lipid bilayer or the requirement for helix formation of the TM segment,[49] it might also be a function of a direct interaction of the TM segment with the protein subunits of the ER translocon during integration in a presumptive protein binding step.

A direct analysis of the molecular environment of TM segments by cross-linking supports a role for the ER translocon in recognition, positioning and release of TM segments during membrane integration. Cross-linking profiles generated by incorporating a probe in three consecutive positions within a TM segment were found to be both sequence-dependent and nonrandom with respect to Sec61α.[50] In this study, specific adducts formed with ER translocon components were retained throughout multiple stages of biosynthesis and beyond the point at which TM segment diffusion into the membrane would be expected if only lipid interactions determined the timing of the partitioning event.[50] It is clear that adduct formation with components of the ER translocon is highly specific to sequence and even context within the nascent chain,[32,50-54] strongly suggesting a role for the ER translocon in regulating the integration process. Furthermore, a posttranslational energy-dependent step required for full membrane integration has been observed for at least one complex polytopic membrane protein.[55]

Thus, whilst the thermodynamic properties of the TM segment are clearly a vital driving force for membrane integration, it is likely that the ER translocon regulates TM segment integration into the membrane via direct protein-protein interactions. The interface at TM2/TM7 of the α-subunit is a logical candidate for any such binding or retention domain within the core ER translocon, given its putative role in the recognition of the secretory protein signal sequence.[14,43,45] However, it may also be that auxiliary translocon components such as TRAM, PAT-10 or the TRAP complex contribute to the process. The analysis of complex, polytopic membrane protein biosynthesis and integration also challenges the simple passive partitioning of TM segments into the lipid bilayer and demands new conceptual descriptions of ER translocon function.

Polytopic Membrane Protein Integration

Since the most salient observation from the detailed studies so far conducted is that TM segment integration in multi-spanning proteins is highly substrate and sequence dependent,[30,32,50-53,56-58] a universally applicable model to describe all polytopic membrane protein integration is inappropriate. Nevertheless, some general principals can be applied. Membrane targeting of the ribosome and nascent chain and its docking at the ER translocon are achieved by the action of SRP and SR in much the same way as for secretory and simple single-spanning membrane proteins. On the resumption of translation at the translocon, the transient α-helix formation of a TM segment is detected at a subtle constriction within the ribosome channel[59] and marks the synthesis of a membrane-spanning region in the protein. Although no specific signalling pathway between the ribosome and the ER translocon has been identified, this event coincides neatly with specific conformational rearrangements at the ER translocon,[60] making direct signal transduction a compelling possibility. As the TM segment enters the aqueous protein-conducting channel it is recognised by the translocon and its translocation into the ER lumen is prevented.[50] Depending on the orientation of the first TM segment, synthesis of the polypeptide sequence C-terminal to the TM segment will either be directly into the lumen of the ER or into the cytosol. In this latter case, a gap between the ribosome and the translocon may be generated to allow the polypeptide chain to be extruded into the cytosol.[41,60] During the synthesis of subsequent TM segments, this pattern will be repeated until the complete protein has been generated and translation terminates. In some exceptional cases, a particularly long cytosolic region between two TM segments may require the retargeting of the ribosome to the membrane by SRP in order to integrate subsequent TM segments correctly.[61] The relative timing of the partitioning for each TM segment into the lipid might be expected to depend on its sequence, length and hydrophobicity as discussed above for a single-spanning membrane protein. However, for polytopic membrane proteins, this appears to be an oversimplification.

In the most elementary model of polytopic membrane protein integration, each TM segment in the polypeptide chain would be synthesised and incorporated into the lipid phase independently and sequentially.[62] Indeed, a detailed study of the aquaporin-4 (AQP4) protein clearly demonstrated that its six TM segments engage the translocon in the order that they are synthesised by the ribosome[52] and sequential entry of TM segments at the ER translocon has also been shown for the opsin protein (Fig. 5).[32,51,57] Moreover, whilst some TM segments within opsin required the synthesis of subsequent segments for displacement from the translocon to occur, others were capable of context-independent integration.[51,57] This is similar to the membrane integration observed for single-spanning membrane proteins, where the exit of the TM segment from the ER translocon may occur prior to the termination of translation at the ribosome.[63]

However, lateral exit from the translocon site for polytopic proteins has also been shown to occur both pairwise[64] and en masse[51,65] and the most striking observation from the cross-linking analyses of AQP4 and opsin biosynthesis is that multiple TM segments are found directly adjacent to translocon components simultaneously (Fig. 5).[5,32,50-53,56] Retention of the TM segments of a complex polypeptide chain at the ER translocon may promote the correct folding and assembly of the protein. Hence, the ER translocon may provide a specific environment that allows the formation of structurally and functionally important ionic interactions between TM segments.[66,67] Interactions between TM segments may also be required for membrane partitioning where proteins contain one or more TM segments of sub-optimal hydrophobicity. In this way, selective TM segment retention at the ER translocon may facilitate an orchestrated assembly of TM segment bundles to generate a structure that can enter the hydrophobic phase of the lipid membrane. Such TM segment cooperation during integration has been observed in an artificial double-spanning membrane protein[56] and also for TM7 of opsin, which has a necessarily compromised hydrophobicity due to the presence of a biologically essential chromophore attachment motif (see Fig. 5).[51,68] Thus, a chaperone-like

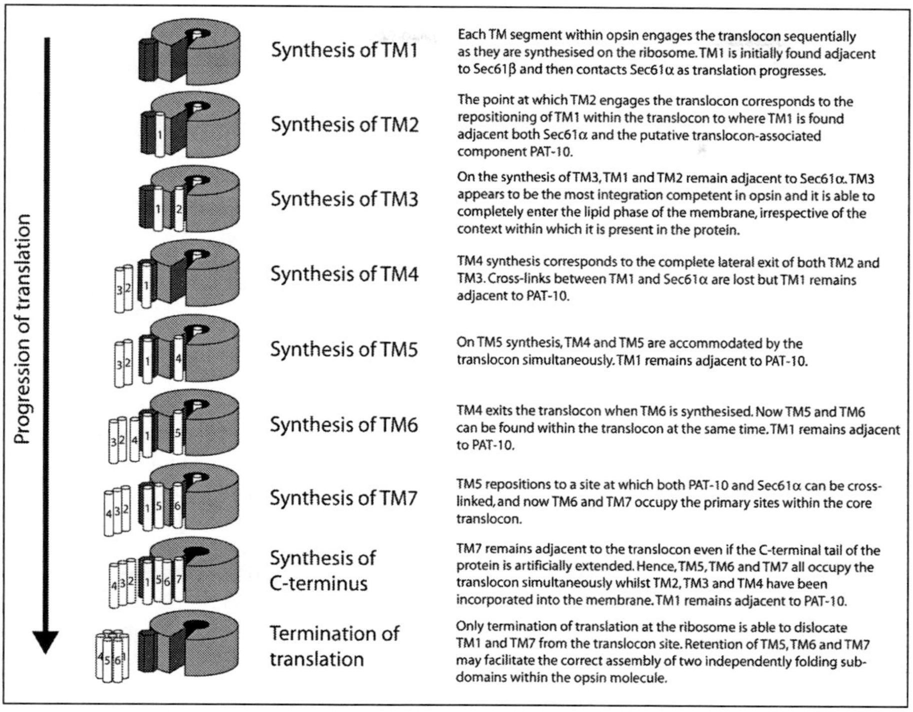

Figure 5. Membrane integration of a model polytopic membrane protein. The membrane integration behaviour of the opsin protein, which has seven TM segments, has been described using careful cross-linking analysis to define the molecular environment encountered by each TM segment as it is integrated into the membrane. The ER translocon is shown with the putative translocon-associated component PAT-10 (shown as a dark grey hexagonal polyhedron) that is found adjacent to opsin during its biosynthesis. The regions comprising TM1-TM4 and TM5-TM7 in opsin represent two independently folding subdomains (see ref. 67) and interestingly both TM1 and TM5 exhibit a pronounced and prolonged association with the ER translocon/PAT-10. Thus, it has been suggested that the translocon may selectively retain specific TM helices in order to allow the orchestrated assembly of complex polytopic membrane proteins (see refs. 32, 51, 57 and 58).

role appears to be fulfilled by the ER translocon during membrane protein biosynthesis and the capacity of the ER translocon to accommodate multiple TM segments simultaneously represents an important mechanism for polytopic membrane protein biosynthesis.

The observations detailed above are difficult to rationalise with the high-resolution structural analysis of the archaeal translocon, which indicates that the pore of the Sec61 complex may be able to accommodate only a single TM helix at a time.[14] Furthermore, by artificially preventing the translocation of both the N- and C-terminal region of a polytopic protein, two hydrophilic regions within the nascent chain were found to span the membrane at the same time.[69] Most interestingly, both of the regions were found to be translocation competent and could be redistributed to the lumen of the ER even if up to seven TM segments were synthesised between them.[69] This might suggest that the hydrophilic regions are accommodated by the translocon pore and would demand an expanded channel within the structure of the translocon.[70] Such an accommodation may be provided by translocon-associated components such as TRAM and the TRAP complex or by oligomerisation of the core Sec61 heterotrimer. Alternatively, a so called front-to-front arrangement of Sec61 complex dimers might provide the basis for a consolidated, enlarged pore, with the two proposed lateral gates facing one another directly (Fig. 4B).[28,29] Cryo-EM analysis of the mammalian ER translocon[24] and fitting of the X-ray structure of the archaeal translocon into electron density maps of the mammalian ER translocon,[23,29] agree that the native arrangement is most likely a dimer of dimers (see Fig. 4A and 4B). On this basis it has also been suggested that the lateral gates of the four Sec61 heterotrimers may all face a central lipid cavity into which membrane proteins may be released during their synthesis (Fig. 4C).[70] In this model, transient retention of the TM segments between the Sec61 complexes may allow for optimised protein folding conditions, whilst the termination of translation and subsequent departure of the ribosome may trigger the conformational rearrangements required to release the protein from within cavity. However, it remains unclear how the known involvement of translocon-associated components may fit into such a model and at present, structural analyses offer little support.

Concluding Remarks

The primary components and some of the mechanistic principals of membrane protein biosynthesis at the ER are now well defined. However, a major target is to reach a consensus on the structure of the ER translocon when it is actively translocating and integrating different classes of protein. Such a versatile complex clearly requires a multifaceted approach to fully describe its activity and the synthesis of both new and existing structural and biochemical tools and model systems will help to achieve these goals. With respect to TM segment partitioning, the availability of quantitative information[46,48] is a huge step forward in understanding the thermodynamic principals of membrane protein biosynthesis. However, as cross-linking studies have clearly demonstrated, further research is needed to define a truly predictive index of TM segment partitioning in both simple and polytopic membrane proteins. The analysis of specific TM sequence motifs that determine TM segment interactions and folding environments may allow better prediction of TM segment behaviour at the ER translocon. Polytopic membrane protein integration remains less well understood than the biosynthesis of single-spanning proteins, but the research conducted to date has already helped to shed light on the workings of the ER translocon. In particular, very few model polytopic proteins have been studied in detail and since those that have been appear to behave very differently at the ER translocon, a universal model for membrane integration is so far incomplete. Hence, understanding how the translocon is able to integrate proteins containing one or more sub-optimal TM segments maybe provide some interesting and important insights into the integration process. Direct analysis of the TM segment itself in the context of the nascent membrane protein, for example by cross-linking and FRET analysis, remains the most revealing approach to describing the integration of complex membrane proteins. Thus, the combination of these techniques with novel interrogative systems will doubtless provide much needed clarity and help to define the general principals of membrane protein integration at the ER translocon.

Acknowledgements

B.C.S.C. is the recipient of BBSRC Doctoral Training Award. Research in the lab of S.H. is supported by funding from the BBSRC, the MRC and the Wellcome Trust. We wish to thank Martin Pool for helpful comments during the preparation of this chapter.

References

1. von Heijne G. The distribution of positively charged residues in bacterial inner membrane proteins correlates with the trans-membrane topology. EMBO J 1986; 5:3021-3027.
2. Walter P, Blobel G. Disassembly and reconstitution of signal recognition particle. Cell 1983; 34:525-533.
3. Keenan RJ, Freymann DM, Stroud RM et al. The signal recognition particle. Annual Review of Biochemistry 2001; 70:755-775.
4. Pool MR. Signal recognition particles in chloroplasts, bacteria, yeast and mammals (review). Mol Membr Biol 2005; 22:3-15.
5. Keenan RJ, Freymann DM, Walter P et al. Crystal structure of the signal sequence binding subunit of the signal recognition particle. Cell 1998; 94:181-191.

6. Lutcke H, High S, Romisch K et al. The methionine-rich domain of the 54 kDa subunit of signal recognition particle is sufficient for the interaction with signal sequences. EMBO J 1992; 11:1543-1551.
7. Walter P, Blobel G. Translocation of proteins across the endoplasmic reticulum III. Signal recognition protein (SRP) causes signal sequence-dependent and site-specific arrest of chain elongation that is released by microsomal membranes. J Cell Biol 1981; 91:557-561.
8. Lakkaraju AK, Mary C, Scherrer A et al. SRP keeps polypeptides translocation-competent by slowing translation to match limiting ER-targeting sites. Cell 2008; 133:440-451.
9. Rabu C, High S. Membrane protein chaperones: a new twist in the tail? Curr Biol 2007; 17:R472-474.
10. Wilkinson BM, Regnacq M, Stirling CJ. Protein translocation across the membrane of the endoplasmic reticulum. The Journal of Membrane Biology 1997; 155:189-197.
11. Egea PF, Stroud RM, Walter P. Targeting proteins to membranes: structure of the signal recognition particle. Curr Opin Struct Biol 2005; 15:213-220.
12. Shan SO, Chandrasekar S, Walter P. Conformational changes in the GTPase modules of the signal reception particle and its receptor drive initiation of protein translocation. J Cell Biol 2007; 178:611-620.
13. Osborne AR, Rapoport TA, van den Berg B. Protein translocation by the sec61/secy channel. Annu Rev Cell Dev Biol 2005; 21:529-550.
14. Van den Berg B, Clemons WM Jr, Collinson I et al. X-ray structure of a protein-conducting channel. Nature 2004; 427:36-44.
15. Harris CR, Silhavy TJ. Mapping an interface of SecY (PrlA) and SecE (PrlG) by using synthetic phenotypes and in vivo cross-linking. J Bacteriol 1999; 181:3438-3444.
16. Li W, Schulman S, Boyd D et al. The plug domain of the secy protein stabilizes the closed state of the translocation channel and maintains a membrane seal. Mol Cell 2007; 26:511-521.
17. Saparov SM, Erlandson K, Cannon K et al. Determining the conductance of the SecY protein translocation channel for small molecules. Mol Cell 2007; 26:501-509.
18. Alder NN, Shen Y, Brodsky JL et al. The molecular mechanisms underlying BiP-mediated gating of the Sec61 translocon of the endoplasmic reticulum. J Cell Biol 2005; 168:389-399.
19. Haigh NG, Johnson AE. A new role for BiP: closing the aqueous translocon pore during protein integration into the ER membrane. J Cell Biol 2002; 156:261-270.
20. Lizak B, Csala M, Benedetti A et al. The translocon and the nonspecific transport of small molecules in the endoplasmic reticulum (Review). Mol Membr Biol 2008; 25:95-101.
21. Cannon KS, Or E, Clemons WM Jr et al. Disulfide bridge formation between SecY and a translocating polypeptide localizes the translocation pore to the center of SecY. J Cell Biol 2005; 169:219-225.
22. Hamman BD, Chen JC, Johnson EE et al. The aqueous pore through the translocon has a diameter of 40-60 A during cotranslational protein translocation at the ER membrane. Cell 1997; 89:535-544.
23. Menetret JF, Hegde RS, Heinrich SU et al. Architecture of the ribosome-channel complex derived from native membranes. J Mol Biol 2005; 348:445-457.
24. Beckmann R, Spahn CM, Eswar N et al. Architecture of the protein-conducting channel associated with the translating 80S ribosome. Cell 2001; 107:361-372.
25. Wirth A, Jung M, Bies C et al. The Sec61p complex is a dynamic precursor activated channel. Mol Cell 2003; 12:261-268.
26. Kalies KU, Rapoport TA, Hartmann E. The beta subunit of the Sec61 complex facilitates cotranslational protein transport and interacts with the signal peptidase during translocation. J Cell Biol 1998; 141:887-894.
27. Osborne AR, Rapoport TA. Protein translocation is mediated by oligomers of the SecY complex with one SecY copy forming the channel. Cell 2007; 129:97-110.
28. Driessen AJ. Cell biology: two pores better than one? Nature 2005; 438:299-300.
29. Mitra K, Schaffitzel C, Shaikh T et al. Structure of the E. coli protein-conducting channel bound to a translating ribosome. Nature 2005; 438:318-324.
30. Sauri A, McCormick PJ, Johnson AE et al. Sec61alpha and TRAM are sequentially adjacent to a nascent viral membrane protein during its ER integration. J Mol Biol 2007; 366:366-374.
31. Heinrich SU, Mothes W, Brunner J et al. The Sec61p complex mediates the integration of a membrane protein by allowing lipid partitioning of the transmembrane domain. Cell 2000; 102:233-244.
32. Meacock SL, Lecomte FJ, Crawshaw SG et al. Different transmembrane domains associate with distinct endoplasmic reticulum components during membrane integration of a polytopic protein. Mol Biol Cell 2002; 13:4114-4129.
33. Voigt S, Jungnickel B, Hartmann E et al. Signal sequence-dependent function of the TRAM protein during early phases of protein transport across the endoplasmic reticulum membrane. J Cell Biol 1996; 134:25-35.
34. Gorlich D, Hartmann E, Prehn S et al. A protein of the endoplasmic reticulum involved early in polypeptide translocation. Nature 1992; 357:47-52.

35. Gorlich D, Rapoport TA. Protein translocation into proteoliposomes reconstituted from purified components of the endoplasmic reticulum membrane. Cell 1993; 75:615-630.
36. Alder NN, Johnson AE. Cotranslational membrane protein biogenesis at the endoplasmic reticulum. J Biol Chem 2004; 279:22787-22790.
37. Tyedmers J, Lerner M, Bies C et al. Homologs of the yeast Sec complex subunits Sec62p and Sec63p are abundant proteins in dog pancreas microsomes. Proc Natl Acad Sci USA 2000; 97:7214-7219.
38. Fons RD, Bogert BA, Hegde RS. Substrate-specific function of the translocon-associated protein complex during translocation across the ER membrane. J Cell Biol 2003; 160:529-539.
39. Martoglio B, Hofmann MW, Brunner J et al. The protein-conducting channel in the membrane of the endoplasmic reticulum is open laterally toward the lipid bilayer. Cell 1995; 81:207-214.
40. High S, Martoglio B, Gorlich D et al. Site-specific photocross-linking reveals that Sec61p and TRAM contact different regions of a membrane-inserted signal sequence. J Biol Chem 1993; 268:26745-26751.
41. Crowley KS, Reinhart GD, Johnson AE. The signal sequence moves through a ribosomal tunnel into a noncytoplasmic aqueous environment at the ER membrane early in translocation. Cell 1993; 73:1101-1115.
42. Jungnickel B, Rapoport TA. A posttargeting signal sequence recognition event in the endoplasmic reticulum membrane. Cell 1995; 82:261-270.
43. Plath K, Mothes W, Wilkinson BM et al. Signal sequence recognition in posttranslational protein transport across the yeast ER membrane. Cell 1998; 94:795-807.
44. Mothes W, Prehn S, Rapoport TA. Systematic probing of the environment of a translocating secretory protein during translocation through the ER membrane. EMBO J 1994; 13:3973-3982.
45. Junne T, Schwede T, Goder V et al. Mutations in the Sec61p channel affecting signal sequence recognition and membrane protein topology. J Biol Chem 2007; 282:33201-33209.
46. Hessa T, Kim H, Bihlmaier K et al. Recognition of transmembrane helices by the endoplasmic reticulum translocon. Nature 2005; 433:377-381.
47. Zhao G, London E. An amino acid "transmembrane tendency" scale that approaches the theoretical limit to accuracy for prediction of transmembrane helices: relationship to biological hydrophobicity. Protein Sci 2006; 15:1987-2001.
48. Hessa T, Meindl-Beinker NM, Bernsel A et al. Molecular code for transmembrane-helix recognition by the Sec61 translocon. Nature 2007; 450:1026-1030.
49. White SH, von Heijne G. Do protein-lipid interactions determine the recognition of transmembrane helices at the ER translocon? Biochem Soc Trans 2005; 33:1012-1015.
50. McCormick PJ, Miao Y, Shao Y et al. Cotranslational protein integration into the ER membrane is mediated by the binding of nascent chains to translocon proteins. Mol Cell 2003; 12:329-341.
51. Ismail N, Crawshaw SG, Cross BC et al. Specific transmembrane segments are selectively delayed at the ER translocon during opsin biogenesis. Biochem J 2008; 411:495-506.
52. Sadlish H, Pitonzo D, Johnson AE et al. Sequential triage of transmembrane segments by Sec61alpha during biogenesis of a native multispanning membrane protein. Nat Struct Mol Biol 2005; 12:870-878.
53. Sauri A, Saksena S, Salgado J et al. Double-spanning plant viral movement protein integration into the endoplasmic reticulum membrane is signal recognition particle-dependent, translocon-mediated and concerted. J Biol Chem 2005; 280:25907-25912.
54. Do H, Falcone D, Lin J et al. The cotranslational integration of membrane proteins into the phospholipid bilayer is a multistep process. Cell 1996; 85:369-378.
55. Oberdorf J, Pitonzo D, Skach WR. An Energy-dependent maturation step is required for release of the cystic fibrosis transmembrane conductance regulator from early endoplasmic reticulum biosynthetic machinery. J Biol Chem 2005; 280:38193-38202.
56. Heinrich SU, Rapoport TA. Cooperation of transmembrane segments during the integration of a double-spanning protein into the ER membrane. EMBO J 2003; 22:3654-3663.
57. Ismail N, Crawshaw SG, High S. Active and passive displacement of transmembrane domains both occur during opsin biogenesis at the Sec61 translocon. J Cell Sci 2006; 119:2826-2836.
58. Laird V, High S. Discrete cross-linking products identified during membrane protein biosynthesis. J Biol Chem 1997; 272:1983-1989.
59. Woolhead CA, McCormick PJ, Johnson AE. Nascent membrane and secretory proteins differ in FRET-detected folding far inside the ribosome and in their exposure to ribosomal proteins. Cell 2004; 116:725-736.
60. Liao S, Lin J, Do H et al. Both lumenal and cytosolic gating of the aqueous ER translocon pore are regulated from inside the ribosome during membrane protein integration. Cell 1997; 90:31-41.
61. Kuroiwa T, Sakaguchi M, Omura T et al. Reinitiation of protein translocation across the endoplasmic reticulum membrane for the topogenesis of multispanning membrane proteins. J Biol Chem 1996; 271:6423-6428.
62. Blobel G. Intracellular protein topogenesis. Proc Natl Acad Sci USA 1980; 77:1496-1500.

63. Mothes W, Heinrich SU, Graf R et al. Molecular mechanism of membrane protein integration into the endoplasmic reticulum. Cell 1997; 89:523-533.
64. Skach WR, Lingappa VR. Amino-terminal assembly of human P-glycoprotein at the endoplasmic reticulum is directed by cooperative actions of two internal sequences. J Biol Chem 1993; 268:23552-23561.
65. Borel AC, Simon SM. Biogenesis of polytopic membrane proteins: membrane segments assemble within translocation channels prior to membrane integration. Cell 1996; 85:379-389.
66. Buck TM, Wagner J, Grund S et al. A novel tripartite motif involved in aquaporin topogenesis, monomer folding and tetramerization. Nat Struct Mol Biol 2007; 14:762-769.
67. Meindl-Beinker NM, Lundin C, Nilsson I et al. Asn- and Asp-mediated interactions between transmembrane helices during translocon-mediated membrane protein assembly. EMBO Rep 2006; 7:1111-1116.
68. Palczewski K. G protein-coupled receptor rhodopsin. Annual review of biochemistry 2006; 75:743-767.
69. Kida Y, Morimoto F, Sakaguchi M. Two translocating hydrophilic segments of a nascent chain span the ER membrane during multispanning protein topogenesis. J Cell Biol 2007; 179:1441-1452.
70. Skach WR. The expanding role of the ER translocon in membrane protein folding. J Cell Biol 2007; 179:1333-1335.

Chapter 8

Membrane Insertion of Tail-Anchored Proteins

Nica Borgese,* Silvia Brambillasca, Marco Righi and Sara Colombo

Abstract

Tail-anchored (TA) proteins are a heterogeneous group of transmembrane proteins with diverse important functions that share an unusual mechanism of insertion into their target membranes. Indeed, because of the C-terminal position of their transmembrane domain they do not have a chance to interact cotranslationally with Signal Recognition Particle and are thus delivered posttranslationally to the ER, to the outer mitochondrial- or to the peroxisomal membrane (as well as to the chloroplast outer envelope in plants). The subsequent insertion involves translocation of the C-terminus, however, TA proteins do not rely on the classical Sec61 translocon nor on the TOM machinery for their integration into the ER and outer mitochondrial membrane respectively. Recent work has demonstrated the existence of different biogenetic pathways for TA proteins. Notably, some are able to efficiently translocate their C-terminus across protein-free bilayers without the participation of any membrane or cytosolic protein, while others require assistance from cytosolic chaperons and membrane receptors. Whether or not assistance is required depends on features of the transmembrane domain. Several intriguing issues in TA protein biogenesis remain to be resolved and will undoubtedly be actively investigated in the coming years. These include the molecular characterization of the ER membrane receptors involved in the assisted pathway and the mechanism by which TA proteins discriminate between the outer mitochondrial membrane and the ER.

Introduction

Tail-anchored (TA) proteins constitute a heterogeneous group of transmembrane proteins defined by the presence of a single transmembrane domain (TMD) close to the C-terminus and an N-terminal domain exposed to the cytosol. The realization that these numerous and functionally diverse proteins differ from other membrane proteins in their mechanisms of targeting to and insertion into membranes has led to a recent surge of interest in TA protein biogenesis.

A recent bioinformatics screen has identified over 400 genes coding for TA proteins in the human genome and confirmed the multiplicity of important roles played by this group of proteins, which range from regulation of apoptosis and of organelle morphology to protein translocation, vesicular transport and lipid trafficking.[1] A similar screen that is currently being carryied out by our group has identified a considerable number of putative TA proteins also in archea and in eubacteria, suggesting that TA proteins are present in all 3 domains of life (Righi and Borgese, unpublished). Table 1 gives a list of TA proteins whose biogenesis has been investigated. The reader can consult the table for information on the single TA proteins discussed in this chapter.

*Corresponding Author: Nica Borgese—CNR Institute for Neuroscience, 20129 Milano, Italy. Email: n.borgese@in.cnr.it

Protein Transport into the Endoplasmic Reticulum, edited by Richard Zimmermann.

Table 1. TA proteins cited in this chapter*

Protein	Function	Localization
Enzymes		
Cytochrome b5	Electron transport	ER
Protein tyrosine phosphatase (PTP)1B	Tyrosine phosphatase involved in the regulation of insulin signalling	ER
SNAREs	Membrane fusion in vesicular transport	
Synaptobrevin/VAMP2		Synaptic vesicles
Syntaxin I		Plasma membrane
S. cerevisiae Sed5		Golgi complex
Components of, or associated with translocation apparatus		
Sec61ß	Component of Sec61 heterotrimer	ER
RAMP4	Associated with translocation site—assists insertion of membrane proteins	ER
Bcl2 family	Regulation of apoptosis	
Bcl2		OMM and ER
Bak		OMM
Bcl-XL		OMM
Adaptors		
S. cerevisiae Fis1	Provides binding site for mitochondrial fission protein Dmn1	OMM
OMP25	Provides binding site for synaptojanin 2A, an inositol 5′ phosphatase	OMM

*Unless otherwise specified, the table refers to mammalian proteins.

Results obtained in the nineties and confirmed by later work (reviewed in refs. 4-6) demonstrated that TA proteins reach their destinations within the secretory pathway by first inserting into the ER and then by classical vesicular traffic.[2,3] In addition to integrating into the ER membrane, TA proteins can also directly insert into outer mitochondrial and peroxisomal membranes and into the chloroplast envelope in plants (reviewed in ref. 6). This chapter will be dedicated principally to the insertion mechanism of ER-targeted TA proteins, but we will briefly address the problem of targeting to other organelles in the last part.

Nearly all membrane proteins targeted to the ER use a Signal Recognition Particle (SRP), Sec61 translocon, cotranslational insertion mechanism (see chapter by Cross and High).[7,8] This is because either an N-terminal signal peptide or the first internal hydrophobic segment is recognized by SRP during ongoing translation, leading to the subsequent well-characterized steps of delivery to and translocation through the heterotrimeric Sec61 protein conducting channel. In the case of TA proteins, the only hydrophobic membrane spanning sequence is so close to the C-terminus (<30 residues) that it remains sequestered within the ribosomal tunnel until release of the terminated polypeptide chain. Thus, TA nascent polypeptides cannot interact with SRP and it is the completed protein that must integrate into the ER bilayer. Although a great deal of progress has been recently made towards unravelling the mechanism of this posttranslational insertion, we are still far away from the wealth of knowledge available on the integration of other membrane proteins.

Assays to Investigate Transmembrane Integration of TA Proteins

One problem that held back the investigation of TA protein biogenesis for a long time was the lack of reliable assays to follow their bona fide transmembrane integration. Indeed, because of the very limited length of the translocated portion of TA proteins, traditional methods based on protection from proteolysis of translocated domains could not be applied. Therefore, the older studies generally used tight, alkali resistant, binding to membranes as a criterion to follow targeting and membrane integration. These assays, however, did not distinguish between transmembrane integration and binding restricted to the cytosolic leaflet of the bilayer and were, in addition, fraught with the problem of a low signal to background ratio.

To overcome the problems associated with these binding assays, our laboratory probed for transmembrane integration of TA substrates using constructs tagged at their C-terminus with a short sequence that provides an N-glycosylation site. The tag allowed us to monitor translocation both by the use of proteolysis protection as well as by utilization of the glycosylation site (Fig. 1).

Recently, another assay, based on the inaccessibility of a single cys residue within the TMD to a membrane-impermeable sulfhydryl-reactive reagent, has been introduced to investigate TA protein insertion.[9,10] While results obtained with this method do not formally demonstrate that

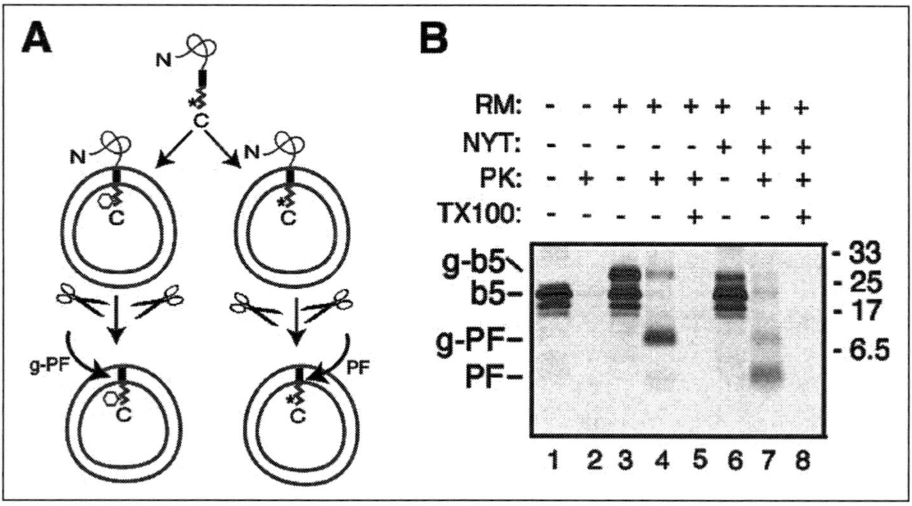

Figure 1. Illustration of protease protection assay for the transmembrane integration of TA proteins. A) schematic representation of the assay. The TA protein under study is equipped with an epitope (grey zig-zag line) immediately downstream to the TMD (black rectangle). The epitope also provides an N-glycosylation consensus sequence (asterisk). After incubation with the TA substrate, vesicles are exposed to protease (scissors). Inserted proteins will generate a PF, which is glycosylated (g-PF) in the case of ER microsomes (left -indicated by hexagon) and unglycosylated in the case of vesicles devoid of oligosaccharyl transferase activity (right). B) Example of results from the assay. SDS-PAGE autoradiography of in vitro tagged cyt b5. After removal of ribosomes, the in vitro translated protein was incubated either without or with rough microsomes (RM), as indicated. The samples were then incubated with protease K (PK), in the presence of Triton-X100 (TX100), as indicated. In some of the samples the tripeptide NYT, a competitive inhibitor of oligosaccharyl transferase, was present during incubation with microsomes. b5 becomes glycosylated after incubation with microsomes (g-b5- lane 3) and generates a glycosylated protected fragment (g-PF) after digestion with PK in the absence of detergent (lane 4). In the presence of NYT, glycosylation is inhibited (lane 6) and an unglycosylated protected fragment (PF) is generated (lane 7). No protection from proteolysis is observed when b5 was not incubated vesicles (lane 2) or if digestion was performed in the presence of detergent (lanes 5 and 8). Panel B is reproduced with permission from reference 11.

the hydrophobic anchor spans the membrane, the assay has the advantage of avoiding addition of an artificial polar sequence to the C-terminus of the investigated protein.

The importance of stringent assays for TA protein transmembrane integration is illustrated by comparison of the results obtained by proteolysis protection with those generated on the basis of binding assays.[11] With the proteolysis protection assay, Brambillasca et al showed that integration of the TA protein cytochrome (cyt) b5 into vesicles is nearly completely blocked by even low concentrations of cholesterol in the bilayer (see paragraph on the unassisted pathway).[11] When the most stringent binding assay available was used (flotation of vesicles in an alkaline sucrose gradient), the inhibition by cholesterol was much less pronounced, indicating that a portion of cyt b5 molecules is tightly bound to, but not integrated in, the cholesterol containing vesicles.

Assisted and Unassisted Insertion of ER-Targeted TA Proteins

There is now a general consensus that at least two pathways operate in TA protein insertion.[6,12] A large body of evidence indicates that the Sec61 translocon is not involved in either pathway. Already in 1995, work based on binding assays, indicated that the TA protein Synaptobrevin/VAMP 2 could tightly bind to proteoliposomes derived from microsomal extracts depleted of Sec61.[2] Subsequent work demonstrated that in vivo translocation of cyt b5's C-terminus in *S. cerevisiae* was unaffected by mutations that inactivate classical co and posttranslational translocation.[13] Moreover, microsomes isolated from similar yeast mutants are fully competent to bind in vitro synthesized TA proteins.[14] Finally, Stefanovic and Hegde have demonstrated protection from proteolysis of Sec61β's C-terminus translocated across translocon-depleted proteoliposomes.[15] Thus, the reported cross-linking between newly synthesized TA substrates and the Sec61 translocon presumably does not reflect a functional association.[16]

Most of the TA substrates used in the cited experiments, although not dependent on the Sec61 translocon, do however require at least one proteinaceous component of the ER (reviewed in ref. 5) We refer to this pathway as the assisted pathway. In contrast, a few substrates (cyt b5, Protein Tyrosine Phosphatase 1B (PTP1B)) can insert into protein-free liposomes or protease-treated microsomes by an unassisted pathway.[17] The use of chimaeric or mutated constructs has led to the conclusion that requirement for or independence from assistance depends entirely on the TMD.[17] Furthermore, rather than a specific sequence, the different behaviour of TA proteins appears to depend in large part on the physicochemical characteristics of the TMD: longer and more hydrophobic TMDs require assistance, while relatively short and moderately hydrophobic TMDs, like the ones of cyt b5 and PTP1B, can do without. We will first discuss each type of insertion separately and then propose a unifying model for the two pathways.

The Unassisted Pathway

This pathway has been characterized using mammalian cyt b5 as substrate,[11,17] however, PTP1B has a behaviour indistinguishable from b5[17] and this is probably also true for other TA proteins with moderately hydrophobic TMDs (for example, Bcl-2).[18,19]

The main features of this pathway can be summarized as follows: (i) a nonassisted substrate is able to translocate its C-terminus across protein-free liposomes made of pure bovine brain phosphatidylcholine (PC) and, based on kinetic analysis, no ER protein plays even a facilitatory role in this process;[11] furthermore, translocation of the in vitro synthesized protein can occur in the near complete absence of nucleotides.[13,17] (ii) After release from the ribosome, b5 interacts with cytosolic proteins, however, these are not required and do not even facilitate its insertion into protein-free bilayers; in the absence of cytosol, b5 self-associates into large aggregates, from which it can be released to the membrane in an energy-independent manner (S. Colombo and N. Borgese, manuscript submitted for publication); (iii) although ER proteins do not play a role in insertion, lipid composition does exert strong effects. PC liposomes containing cholesterol at levels only slightly higher than those of the ER are incompetent for b5 and PTP1B integration; furthermore, loading ER microsomes with the sterol completely abolishes cyt b5 insertion.[11,17] The sharp inhibition exerted by cholesterol could partly explain the inability of unassisted TA proteins to insert into compartments of the secretory pathway downstream to the ER. iv) The nature of the translocated polar sequence seems to have little influence on the insertion process. Indeed, Brambillasca et al appended different polar sequences to cyt b5's C-terminus and observed that domains as long as 85 residues could be posttranslationally translocated across protein-free PC liposomes as efficiently as across microsomes, independently from their sequence and charge.[17] This observation revealed a previously unsuspected permissivity of fluid lipid bilayers.

The obvious question that arises is the in vivo relevance of the results obtained in a cell-free system. An example that comes to mind is the insertion of the phage coat proteins M13 and Pf3 into lipid bilayers. Although capable of inserting into protein-free liposomes in vitro,[20,21] in vivo they both require the bacterial inner membrane protein YidC.[22,23] However, a comparison of the insertion efficiency of Pf3 into protein-free liposomes and into proteoliposomes reconstituted with YidC demonstrated an up to 4 fold stimulation by YidC also in vitro.[24] In contrast, b5 inserts with indistinguishable efficiency into protein-free liposomes and microsomes, strongly suggesting that it can translocate its C-terminus across bilayers without help from proteins also in vivo. Similar unassisted insertion mechanisms may play a role in the biogenesis also of other organelles, such as the outer mitochondrial membrane (OMM) and chloroplast thylakoid membranes.[25,26]

The finding that a group of TA proteins, which in the cell is specifically localized to one subcellular compartment, can spontaneously insert into protein-free bilayers raises questions as to how they achieve specific targeting in vivo. This problem will be discussed later in the chapter.

The Assisted Pathway

An important advance in unravelling the assisted pathway has come from the discovery of a key cytosolic player, an ATPase previously not suspected to participate in protein targeting.[15] The ATPase is homologous to a bacterial protein involved in heavy metal transport (including arsenite, hence its name, ArsA)[27] and on this basis was previously named Asna1 in mammals and Arr4 in *Saccharomyces*.[28-30] However, its role in metal transport in eukaryotes has never been demonstrated.

Stefanovic and Hegde demonstrated that Asna1 is a major crosslinking partner of a number of TA proteins translated in the rabbit reticulocye lysate, including Synaptobrevin/VAMP 2, Sec61ß, Syntaxin1 (Table 1);[15] the former two are well known substrates of the assisted pathway.[2,16,19] Based on sucrose gradient analysis, Asna1 appears to be part of a larger complex. The complex between the TA protein and Asna1 is an intermediate in the delivery pathway to the ER membrane, because the TA protein is released from Asna1 in an ATP-dependent manner upon its insertion into microsomes. A mutant of Asna1 predicted to have impaired ATPase activity behaved as a dominant negative when added to the translation mix, inhibiting the insertion of the three investigated TA substrates but being quite ineffective on cyt b5 insertion. Thus, Asna1 appears to be part of a major pathway of TA protein delivery to the ER membrane and was accordingly renamed TRC40 (TMD Recognition Complex subunit of 40 kDa). Notably, TRC40/Asna1 recognizes hydrophobic stretches in a particular context, such as the TMD of TA proteins, but does not interact with the TMDs of proteins that are normally delivered to the ER by the cotranslational pathway. Recently, Favaloro et al confirmed the involvement of Asna1/TRC40 in the delivery of TA proteins to the ER, demonstrating its participation in the insertion of RAMP4, a TA protein associated with the translocation site (Table 1).[31]

Recent studies have extended the significance of the results obtained in vitro in mammalian systems to yeast cells, showing a role for the TRC40/Asna1 *S. cerevisiae* orthologue, GET3/Arr4, in TA protein delivery to the ER both in vivo and in vitro.[32,33] The yeast protein exists in a complex with two ER membrane receptors,[32,34] whose deletion causes defects in Golgi to ER traffic. On this basis, the complex was called GET (Golgi to ER traffic),[32] with Arr4 (the Asna1/TRC40 orthologue) renamed GET3 and the two ER membrane proteins GET1 and 2. Of the membrane proteins of the complex, only GET1 has a clear mammalian homolog. Subsequently, the group discovered that the trafficking defect observed in GET deletion strains can be explained by the defective membrane insertion of Sed5, a Golgi SNARE involved in trafficking between the

ER and the Golgi (Table I).[33] Furthermore, the insertion defect is not limited to Sed5 but affects a number of other investigated TA proteins. Defective insertion of TA proteins with widely differing functions can explain the pleiotropic effects of GET deletions,[33] especially considering that nearly all SNAREs are tail-anchored.

Although the discovery of the involvement of the TRC40/GET complex in TA protein delivery to the ER represents a major breakthrough in the field, various observations suggest that additional pathways exist. GET3 deletion in yeast has mild phenotypic effects,[32] suggesting that enough essential TA proteins can reach the ER in its absence as to guarantee cell survival. GET1/2 deletion has stronger effects and TA proteins remain trapped in GET3 containing cytosolic aggregates in GET1/2-deleted cells.[33] Very interestingly, the effects of GET1/2 deletion are mitigated by the additional deletion of GET3, again indicating that TA proteins can reach the ER by alternative pathways. In line with this idea, Abell et al have shown an unusual posttranslational involvement of SRP in TA protein insertion[35] and, more recently, a strong stimulatory effect of heat shock proteins Hsc70 and Hsp40.[36] In addition, it is likely that, whereas cyt b5 and PTP1B represent extreme examples of TA proteins capable of very efficient unassisted insertion, also other TA proteins may use the unassisted pathway, albeit with lower efficiency. The elucidation of the interplay between and the regulation of, the various pathways for TA protein delivery is an important future goal pursued in our and other laboratories.

Does Tail-Anchored Protein Insertion Need an Insertase?

The discovery that at least some TA proteins can translocate their C-terminus across protein-free bilayers raises the question as to whether proteins that follow the assisted pathway carry out the translocation step in the same way. We hypothesize that the need for, or independence from, assistance can be rationalized simply on the basis of different requirements for maintenance of an insertion-competent state of the TA protein after its release from the ribosome. It is known that the poor water solubility of hydrophobic peptides constitutes a major problem for their assembly into preformed lipid bilayers.[37,38] Similarly, TA proteins with very hydrophobic TMDs are probably rapidly sequestered into insoluble aggregates unless assisted by TRC40/GET3 or some other chaperone. The role for ER protein(s) in the insertion process could then be limited to the ATP-dependent release of the TA substrate from its chaperone. In contrast, TA proteins with TMDs of limited hydrophobicity may be endowed with a certain degree of water solubility, which would allow them to directly access the bilayer. Since the transmembrane integration of unassisted TA proteins does not require an insertase, we see no reason to postulate the existence of one to assist the insertion of TA proteins with more hydrophobic TMDs and suggest instead that the final integration step into the bilayer is shared by all TA proteins, as illustrated in Figure 2. The molecular dissection of the entire pathway of TA protein biogenesis will undoubtedly resolve the "insertase" question.

The Targeting Problem

As discussed in the Introduction to this chapter, in eukaryotic cells, TA proteins are present on all membranes abutting the cytosol, including those of the secretory pathway, the OMM, the peroxisomal membrane and, in plant, the outer envelope of the chloroplast. The pathway they follow to reach peroxisomes is controversial, with evidence in favor both of their initial insertion into the ER or specialized subdomains thereof, followed by inclusion into nascent peroxisomes that form from the ER[39-41] and of direct integration into the peroxisomal membrane.[42,43] In the case of mitochondria instead there is clear evidence that transport occurs directly from the cytosol.[44] The discussion here will be limited to the problem of how TA proteins distinguish between the OMM and the ER.

There is general agreement that the tail region of TA proteins contains the main targeting information for discrimination between ER and MOM (reviewed in refs. 5 and 6), although there are a few cases in which information in the cytosolic domain is essential.[45,46] Another caveat is that targeting in yeast may not always recapitulate the process occurring in higher eukaryotes.[47] With

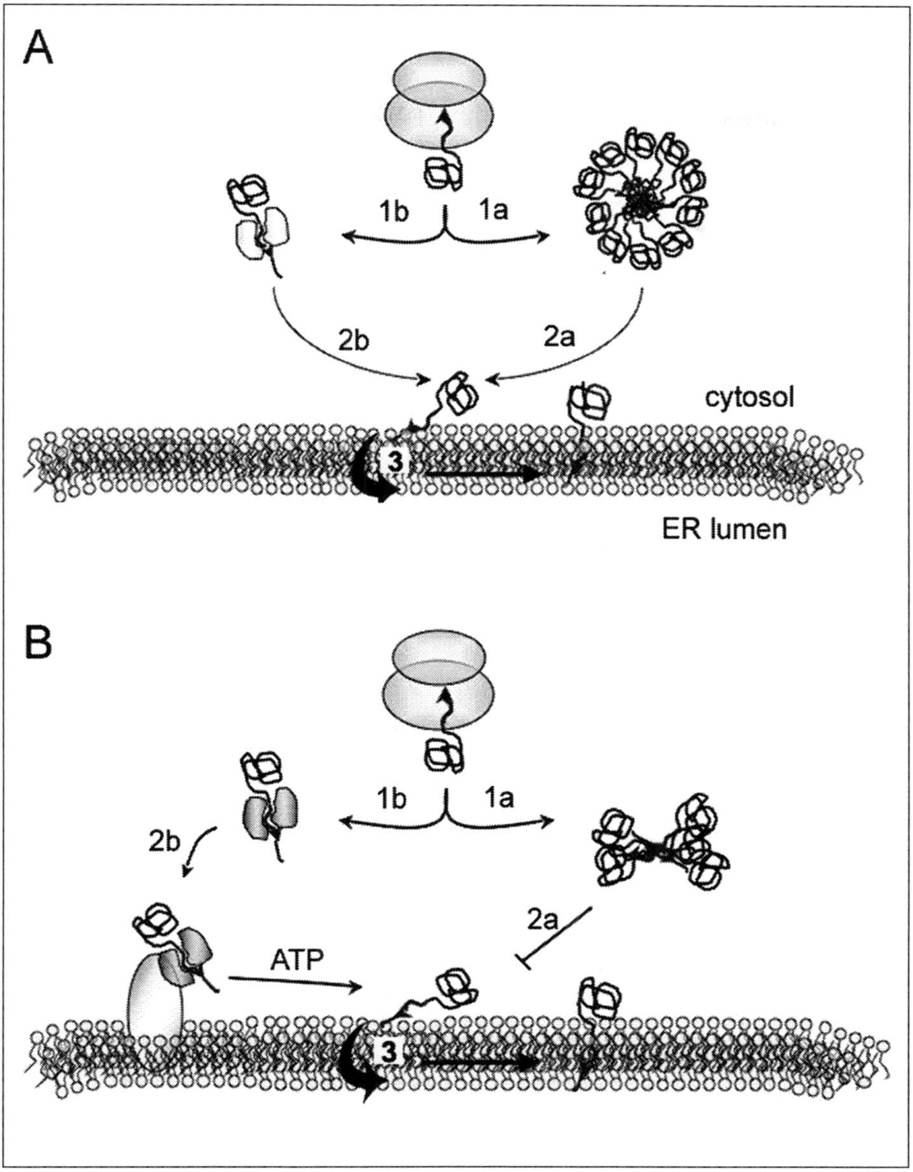

Figure 2. Model of unassisted (A) and assisted (B) insertion. A) Unassisted insertion: after release from the ribosome, the TA protein interacts via its TMD with cytosolic proteins (1b) from which it can be released to the membrane in an energy-independent step (2b), followed by spontaneous translocation of the TMD across the bilayer (3). In the absence of any cytosolic component, the TA protein self-associates via the TMD (1a); these aggregates are in equilibrium with a translocation-competent form (2a), which then translocates across the bilayer (3). B) Assisted insertion: in the absence of cytosolic proteins, these TA proteins self-associate via the TMD to form aggregates (1a), from which release to the membrane is not possible (2a). Dedicated chaperones (TRC40, see text) maintain the TA substrate in an insertion-competent state (1b). Release from the chaperones occurs thanks to an ER receptor (2b) and ATP. The final translocation step (3) occurs directly across lipids, as in the case of unassisted insertion.

these limitations in mind, the ER appears to have broad specificity for tail anchors, allowing the insertion of TMDs of different length and hydrophobicity and with differently charged flanking regions. Instead, although discrete targeting sequences are not involved, distinct physicochemical features of the tail region, i.e., moderate hydrophobicity of the TMD and positive charge in the polar flanking regions, are required for mitochondrial targeting. As recently illustrated, however, although these features on the average clearly distinguish the population of mitochondrially targeted tails from their ER counterparts and although deletion of these properties by site directed mutation results in loss of correct targeting, there is quite some overlap in hydrophobicity and positive charge between mitochondrially and ER targeted TA proteins, so that these features alone cannot explain the specific in vivo targeting which is observed in most cases.[6] A recent report has invoked the propensity for α-helix formation as a third TMD feature that favors mitochondrial targeting.[48] The generality of this interesting observation deserves to be tested on a larger number of artificial and natural sequences with different targeting specificity.

If there are still open questions on the mitochondrial targeting signals of TA proteins, even less is known about the cellular machinery that interprets these signals. It was thought that the outer membrane machinery (TOM complex), which is responsible for the import of precursors directed to essentially all mitochondrial subcompartments, was used also by TA proteins.[49,50] However, a study of the targeting of three in vitro synthesized TA proteins (two members of the bcl2 family and OMP25, a receptor for a phosphatidylinositol bis phosphate phosphatase) to mitochondria in digitonin-permeabilized cultured cells silenced for TOM components, has ruled out the involvement of this complex in TA protein targeting, at least in the mammalian system.[43] Interestingly, a recent study has shown that the yeast OMM protein Fis1 can insert into protein-free lipid vesicles.[10] Difficult to reconcile with this result is the observation that the three mitochondrially targeted TA proteins investigated by Setoguchi et al compete with each other for limiting binding sites.[43] This suggests the possibility that also for mitochondrial TA proteins an assisted and a nonassisted pathway co-exist. As far as the role of cytosolic proteins in the delivery of TA proteins to mitochondria is concerned, the available evidence suggests that none are required.[43,51] Since the TMDs of mitochondrially targeted TA are only moderately hydrophobic, it may not be surprising that they do not need to be chaperoned.

Regardless of the nature of the mitochondrial machinery involved in TA protein recognition/insertion, it is clear that in many cases discrimination between the ER and the OMM occurs by a competitive mechanism. This is illustrated by the recent observation that a subset of TA proteins in *Δget1/Δget2 Saccharomyces* strains (see preceding sections of this chapter) are mistarged to mitochondria.[33] This not only highlights an important role for the GET system in ensuring correct ER targeting, but also illustrates how TA proteins that are normally delivered to the ER are fully competent for insertion into the OMM and that normal in vivo targeting is presumably based on kinetic partitioning. A symmetric example is offered by cyt b5 targeting. Mutation of the C-terminal polar domain of the ER targeted protein to contain a net positive charge results in a protein (b5-RR) that is specifically targeted to mitochondria in vivo. Yet, in vitro, this mutant inserts into ER microsomes as efficiently as its wild-type counterpart.[44] While it is straightforward to envisage competition between cytosolic targeting factors and between membrane receptors of different organelles, it is much more difficult to imagine how specificity is achieved in the case of TA proteins competent for unassisted insertion. Differences in lipid composition of the OMM and ER bilayer and/or interaction with chaperones which are not required for insertion but which influence targeting may both be involved in this delicately balanced and potentially modulatable process.

Conclusion and Perspectives

The functional importance and unusual biogenesis of TA proteins were recognized about 15 years ago[52] and, since then, increasing efforts have been dedicated to the investigation of the post-translational mechanisms underlying their insertion into the correct target membrane. The recent characterization of the different delivery pathways used by TA proteins and the identification of the TRC40/GET complex constitute a quantum leap in the level of our understanding of these processes. The stage now seems set to address and resolve the outstanding questions on TA protein biogenesis, which, in our minds, are: the identification of the TRC40/Asna1 receptor in higher eukaryotes, the resolution of the question as to whether or not an insertase is involved in the transmembrane integration of assisted TA proteins, the identification of the mitochondrial machinery involved in TA protein insertion and the mechanisms by which TA proteins discriminate between the ER and the OMM.

Acknowledgements

The authors wish to acknowledge support from Telethon (Grants GP04129 and GGP07010 to N.B.). S.C. is supported by a research contract awarded by the University of Milan.

References

1. Kalbfleisch T, Cambon A, Wattenberg BW. A bioinformatics approach to identifying tail-anchored proteins in the human genome. Traffic 2007; 8:1687-1694.
2. Kutay U, Ahnert-Hilgen G, Hartmann E et al. Transport route for synaptobrevin via a novel pathway of insertion into the endoplasmic reticulum membrane. EMBO J 1995; 14:217-223.
3. Jäntti J, Keränen S, Toikkanen J et al. Membrane insertion and intracellular transport of yeast syntaxin Sso2p in mammalian cells. J Cell Science 1994; 107:3623-3633.
4. Wattenberg B, Lithgow T. Targeting of C-terminal (tail)-anchored proteins: understanding how cytoplasmic activities are anchored to intracellular membranes. Traffic 2001; 2:66-71.
5. Borgese N, Colombo S, Pedrazzini E. The tale of tail-anchored proteins: coming from the cytosol and looking for a membrane. J Cell Biol 2003; 161:1013-1019.
6. Borgese N, Brambillasca S, Colombo S. How tails guide tail-anchored proteins to their destinations. Curr Opin Cell Biol 2007; 19:368-375.
7. Higy M, Junne T, Spiess M. Topogenesis of membrane proteins at the endoplasmic reticulum. Biochemistry 2004; 43:12716-12722.
8. Rapoport TA. Protein translocation across the eukaryotic endoplasmic reticulum and bacterial plasma membranes. Nature 2007; 450:663-669.
9. Kim PK, Annis MG, Dlugosz PJ et al. During apoptosis bcl-2 changes membrane topology at both the endoplasmic reticulum and mitochondria. Mol Cell 2004; 14:523-529.
10. Kemper C, Habib SJ, Engl G et al. Integration of tail-anchored proteins into the mitochondrial outer membrane does not require any known import components. J Cell Sci 2008; 121:1990-1998
11. Brambillasca S, Yabal M, Soffientini P et al. Transmembrane topogenesis of a tail-anchored protein is modulated by membrane lipid composition. EMBO J 2005; 24:2533-2542.
12. High S, Abell BM. Tail-anchored protein biosynthesis at the endoplasmic reticulum: the same but different. Biochem Soc Trans 2004; 32:659-662.
13. Yabal M, Brambillasca S, Soffientini P et al. Translocation of the C terminus of a tail-anchored protein across the endoplasmic reticulum membrane in yeast mutants defective in signal peptide-driven translocation. J Biol Chem 2003; 278:3489-3496.
14. Steel GJ, Brownsword J, Stirling CJ. Tail-anchored protein insertion into yeast ER requires a novel posttranslational mechanism which is independent of the SEC machinery. Biochemistry 2002; 41:11914-11920.
15. Stefanovic S, Hegde RS. Identification of a targeting factor for posttranslational membrane protein insertion into the ER. Cell 2007; 128:1147-1159.
16. Abell BM, Jung M, Oliver JD et al. Tail-anchored and signal-anchored proteins utilise overlapping pathways during membrane insertion. J Biol Chem 2003; 278:5669-5678.
17. Brambillasca S, Yabal M, Makarow M et al. Unassisted translocation of large polypeptide domains across phospholipid bilayers. J Cell Biol 2006; 175:767-777.
18. Janiak F, Leber B, Andrews DW. Assembly of Bcl-2 into microsomal and outer mitochondrial membranes. J Biol Chem 1994; 269:9842-9849.
19. Kim PK, Janiak-Spens F, Trimble WS et al. Evidence for multiple mechanisms for membrane binding and integration via carboxyl-terminal insertion sequences. Biochemistry 1997; 36:8873-8882.
20. Geller BL, Wickner W. M13 procoat inserts into liposomes in the absence of other membrane proteins. J Biol Chem 1985; 260:13281-13285.
21. Ridder AN, Morein S, Stam JG et al. Analysis of the role of interfacial tryptophan residues in controlling the topology of membrane proteins. Biochemistry 2000; 39:6521-6528.
22. Samuelson JC, Chen M, Jiang F et al. YidC mediates membrane protein insertion in bacteria. Nature 2000; 406:575-577.

23. Chen M, Samuelson JC, Jiang F et al. Direct interaction of YidC with the Sec-independent Pf3 coat protein during its membrane protein insertion. J Biol Chem 2002; 277:7670-7675.
24. Serek J, Bauer-Manz G, Struhalla G et al. Escherichia coli YidC is a membrane insertase for sec-independent proteins. EMBO J 2004; 23:294-301.
25. Kemper C, Habib SJ, Engl G et al. Integration of tail-anchored proteins into the mitochondrial outer membrane does not require any known import components. J Cell Sci 2008; 121:1990-1998.
26. Di Cola A, Klostermann E, Robinson C. The complexity of pathways for protein import into thylakoids: it's not easy being green. Biochem Soc Trans 2005; 33:1024-1027.
27. Rosen BP, Hsu CM, Karkaria CE et al. A plasmid-encoded anion-translocating ATPase. Biochim Biophys Acta 1990; 1018:203-205.
28. Kurdi-Haidar B, Aebi S, Heath D et al. Isolation of the ATP-binding human homolog of the arsA component of the bacterial arsenite transporter. Genomics 1996; 36:486-491.
29. Bhattacharjee H, Ho YS, Rosen BP. Genomic organization and chromosomal localization of the Asna1 gene, a mouse homologue of a bacterial arsenic-translocating ATPase gene. Gene 2001; 272:291-299.
30. Shen J, Hsu CM, Kang BK et al. The saccharomyces cerevisiae Arr4p is involved in metal and heat tolerance. Biometals 2003; 16:369-378.
31. Favaloro V, Spasic M, Schwappach B et al. Distinct targeting pathways for the membrane insertion of tail-anchored (TA) proteins. J Cell Sci 2008; 121:1832-1840.
32. Schuldiner M, Collins SR, Thompson NJ et al. Exploration of the function and organization of the yeast early secretory pathway through an epistatic miniarray profile. Cell 2005; 123:507-519.
33. Schuldiner M, Metz J, Schmid V et al. The GET complex mediates insertion of tail-anchored proteins into the ER membrane. Cell 2008; 134:634-645.
34. Auld KL, Hitchcock AL, Doherty HK et al. The conserved ATPase get3/arr4 modulates the activity of membrane-associated proteins in saccharomyces cerevisiae. Genetics 2006; 174:215-227.
35. Abell BM, Pool MR, Schlenker O et al. Signal recognition particle mediates posttranslational targeting in eukaryotes. EMBO J 2004; 23:2755-2764.
36. Abell BM, Rabu C, Leznicki P et al. Posttranslational integration of tail-anchored proteins is facilitated by defined molecular chaperones. J Cell Sci 2007; 120:1743-1751
37. Hunt JF, Rath P, Rothschild KJ et al. Spontaneous, pH-dependent membrane insertion of a transbilayer alpha-helix. Biochemistry 1997; 36:15177-15192.
38. Reshetnyak YK, Andreev OA, Lehnert U et al. Translocation of molecules into cells by pH-dependent insertion of a transmembrane helix. Proc Natl Acad Sci USA 2006; 103:6460-6465.
39. Elgersma Y, Kwast L, van den Berg M et al. Overexpression of Pex15p, a phosphorylated peroxisomal integral membrane protein required for peroxisome assembly in S. cerevisiae, causes proliferation of the endoplasmic reticulum membrane. EMBO J 1997; 16:7326-7341.
40. Mullen RT, Trelease RN. The sorting signals for peroxisomal membrane-bound ascorbate peroxidase are within its C-terminal tail. J Biol Chem 2000; 275:16337-16344.
41. Hoepfner D, Schildknegt D, Braakman I et al. Contribution of the endoplasmic reticulum to peroxisome formation. Cell 2005; 122:85-95.
42. Halbach A, Landgraf C, Lorenzen S et al. Targeting of the tail-anchored peroxisomal membrane proteins PEX26 and PEX15 occurs through C-terminal PEX19-binding sites. J Cell Sci 2006; 119:2508-2517.
43. Setoguchi K, Otera H, Mihara K. Cytosolic factor- and TOM-independent import of C-tail-anchored mitochondrial outer membrane proteins. EMBO J 2006; 25:5635-5647.
44. Borgese N, Gazzoni I, Barberi M et al. Targeting of a tail-anchored protein to endoplasmic reticulum and mitochondrial outer membrane by independent but competing pathways. Mol Biol Cell 2001; 12:2482-2496.
45. Shirane M, Nakayma KI. Inherent calcineurin inhibitor FKBP38 targets Bcl-2 to mitochondria and inhibits apoptosis. Nature Cell Biol 2003; 5:28-37.
46. Bellot G, Cartron PF, Er E et al. TOM22, a core component of the mitochondria outer membrane protein translocation pore, is a mitochondrial receptor for the proapoptotic protein Bax. Cell Death Differ 2007; 14:785-794.
47. Henderson MP, Hwang YT, Dyer JM et al. The C-terminus of cytochrome b5 confers endoplasmic reticulum specificity by preventing spontaneous insertion into membranes. Biochem J 2007; 401:701-709.
48. Wattenberg BW, Clark D, Brock S. An artificial mitochondrial tail signal/anchor sequence confirms a requirement for moderate hydrophobicity for targeting. Biosci Rep 2007; 27:385-401.
49. Schleiff E, Shore GC, Goping IS. Human mitochondrial import receptor, tom20p. Use of glutathione to reveal specific interactions between tom20-glutathione S-transferase and mitochondrial precursor proteins. FEBS Lett 1997; 404:314-318.

50. Motz C, Martin H, Krimmer T et al. Bcl-2 and porin follow different pathways of TOM-dependent insertion into the mitochondrial outer membrane. J Mol Biol 2002; 323:729-738.
51. Lan L, Isenmann S, Wattenberg BW. Targeting and insertion of C-terminally anchored proteins to the mitochondrial outer membrane is specific and saturable but does not strictly require ATP or molecular chaperones. Biochem J 2000; 349:611-621.
52. Kutay U, Hartmann E, Rapoport TA. A class of membrane proteins with C-terminal anchor. Trends Cell Biol 1993; 3:72-75.

CHAPTER 9

Human Pathologies of Protein Transport into the Endoplasmic Reticulum

Linda Müller, Bernd Wullich and Richard Zimmermann*

Abstract

Protein transport into the endoplasmic reticulum (ER) involves a protein translocase that resides in the ER membrane and is highly conserved between lower and higher eukaryotes. The Sec61 complex forms the central component of this translocase, providing a path for the newly synthesized polypeptide on its way into the membrane or the lumen of the ER. The membrane proteins Sec62 and Sec63, plus the molecular chaperone BiP, are additional subunits of the protein translocase and couple translocation to the hydrolysis of ATP. In humans, ERj1 appears to provide a functional alternative to the Sec62/Sec63 complex. Furthermore, Sil1 and Grp170 are alternatively acting nucleotide exchange factors for BiP. The essential character of BiP was emphasized by the recent discovery that inactivation of human BiP by a bacterial cytotoxin causes life-threatening complications, such as haemolytic uraemic syndrome. On the other hand, genetic work has identified mutations in the human *SEC63* gene as one cause of autosomal dominant polycystic liver disease and mutations in the human *SIL1* gene as one cause for a neurodegenerative disorder. Thus, redundant components of the protein translocation machinery of the ER are targets of genomic alterations that can cause nonlethal human diseases.

Introduction

Protein transport into the endoplasmic reticulum (ER) is the decisive initial step in protein secretion of eukaryotic cells.[1,2] This process is also critical in the biogenesis of proteins of the various organelles that are involved in endo- and exocytosis. Protein transport into the ER involves amino-terminal signal peptides in the precursor proteins (see chapter by Kapp et al) and a highly conserved transport machinery. In *Homo sapiens* as well as in *Saccharomyces cerevisiae,* this machinery consists of proteins and protein complexes in the cytosol and in the ER (Table 1, Fig. 1) (for the mammalian components, see chapter by Dudek et al, respectively; for the yeast components, see chapter by Wilkinson and Stirling).

The cytosolic transport components are dedicated to posttranslational (heat shock proteins, Hsp) or cotranslational (signal recognition particle, SRP) transport. In posttranslational transport, fully synthesized precursor polypeptides are transported with the help of cytosolic molecular chaperones (Hsp70 and Hsp40).[3] The chaperones keep the precursor polypeptides soluble and competent for interaction with the protein translocase in the ER membrane. In cotranslational transport, nascent precursor polypeptides are transported with the help of SRP and the SRP receptor (SR) on the ER surface. SRP binds to nascent precursor polypeptides as their signal peptides

*Corresponding Author: Richard Zimmermann—Medical Biochemistry and Molecular Biology, Saarland University, D-66421 Homburg, Germany. Email: bcrzim@uks.eu

Protein Transport into the Endoplasmic Reticulum, edited by Richard Zimmermann

Table 1. Components of ER-protein translocase as present in human and yeast cells. Alternative names are given in parentheses.

Component	*H. sapiens*	Related Human Disease	*S. cerevisiae*	In *S. cerevisiae* Essential
Sec61-Complex	Sec61α1		Sec61p	+
	Sec61β		Sbh1p	–
	Sec61γ		Sss1p	+
Alternative 61	Sec61α2		Ssh1p	–
	–		Sbh2p	–
	–		Sss1p	+
Sec62-63-Complex	Sec62(TLOC1)		Sec62p	+
	Sec63(ERj2)	Polycystic liver disease	Sec63p (Ptl1p, Npl1p)	+
	–		Sec66p (Sec71p)	–
	–		Sec67p (Sec72p)	–
Alternative 62/63	ERj1 (Mtj1, DNAJC 1)		–	
Trans-Acting Chaperone	BiP (Grp78, HspA5)	Haemolytic uraemic syndrome	Kar2p	+
Nucleotide Exchange Factors	Grp170 (Orp150, HYOU1)		Lhs1p (Ssi1p, Cer1p)	–
	Sil1 (BAP)	Marinesco-Sjögren syndrome	Sil1p	–

emerge from the translating ribosome. This interaction attenuates protein synthesis, allowing the ribosome/nascent chain complex with its attached SRP to bind to SR at the ER membrane. Thus, SRP facilitates ER targeting and simultaneously acts as a molecular chaperone for the nascent polypeptide. Depending on the concentration of rough ER in the particular cell type, a significant fraction of polypeptides may be synthesized on ribosomes that are continuously attached to the ER membrane (see chapter by Nicchitta and Stephens). In this case, SRP and cytosolic chaperones are not required for translocation. Therefore, nascent polypeptides that lack a signal peptide for ER targeting are recognized by the nascent chain associated complex and their synthesis is completed on free ribosomes.[4]

Nascent or fully synthesized precursor polypeptides are integrated into or translocated across the ER membrane by the protein translocase. The central component of this protein translocase is the Sec61 complex, consisting of α-, β- and γ-subunits (Fig. 1). In *S. cerevisiae,* there is an alternative Sec61 complex that has the γ-subunit in common with the major complex but contains homologues of the α- and β-subunits (Table 1). Similarly, two isoforms of Sec61α exist in *H. sapiens*. The Sec61 complex acts as a signal peptide receptor and a ribosome receptor (in cotranslational transport) and forms an aqueous pore for the membrane insertion or translocation of the precursor polypeptide (see chapter by Wagner et al, respectively). The Sec61 complex associates with additional proteins and complexes to form the complete protein translocase (see chapters by Aronson and Snapp and Dudek et al, respectively).

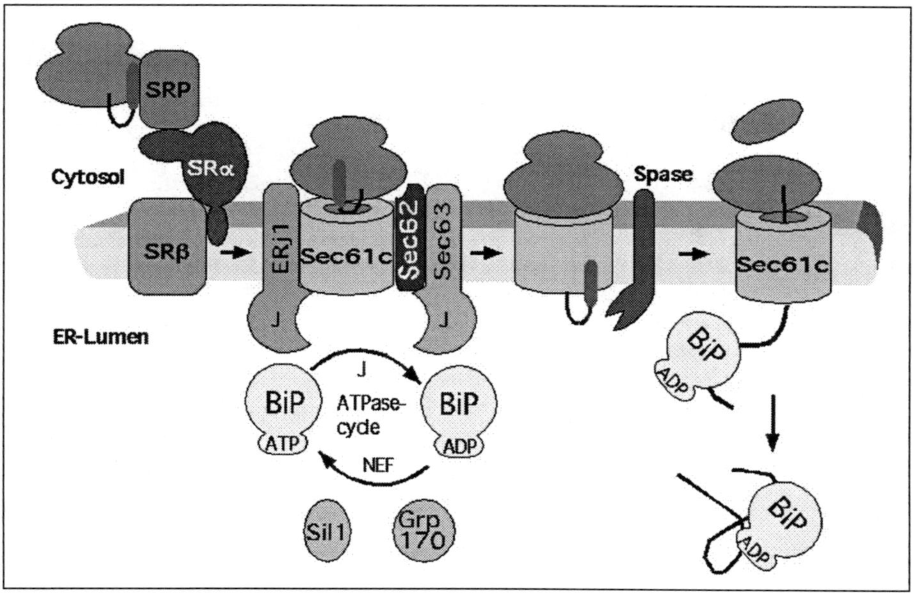

Figure 1. Cotranslational protein translocation across the ER membrane. Cotranslational membrane transport is depicted for precursor polypeptides with amino-terminal signal peptides. The figure was adapted from reference 33.

In both *S. cerevisiae* and *H. sapiens*, cotranslational protein transport into the ER involves the major Hsp70-type molecular chaperone of the ER lumen (Kar2p in yeast;[5] BiP in humans[6]), their Hsp40-type cochaperone (Sec63p in yeast;[7] Sec63 or ERj1 in humans[8-10]) and their nucleotide exchange factors (Sil1p and Lhs1p in yeast;[11,12] Sil1 and Grp170 in humans[13,14]) (Fig. 1 and Table 1). *KAR2* and *SEC63* are essential genes in yeast. In contrast, *LHS1* and *SIL1* are nonessential genes in yeast.[11] However, (i) simultaneous deletion of the latter two genes results in synthetic lethality; (ii) deletion of *LHS1* results in accumulation of presecretory proteins in the cytosol that can be suppressed by overexpression of *SIL1*; and (iii) conditional *SEC63* and *KAR2* mutants and *LHS1/SIL1* double mutants accumulate precursors of secretory proteins in the cytosol at the restrictive temperature.[11] We note that the Hsp70s and their cochaperones and nucleotide exchange factors are also involved in protein folding and protein assembly in the ER and at least BiP and Kar2p are also involved in export of mis-folded polypeptides from the ER for degradation by the proteasome (known as ER-associated protein degradation, or ERAD).[15]

The ER-lumenal Hsp70 BiP performs a classical Hsp70 reaction cycle together with its cochaperones, the ER-lumenal J-domain of Sec63 and its alternative nucleotide exchange factors Sil1 and Grp170 (see chapter by Dudek et al). The J-domain of Sec63 recruits BiP to the Sec61 complex and, by stimulating ATP hydrolysis, activates BiP for interaction with its substrates. Activated BiP has several functions in protein translocation: (i) it is involved in the insertion of precursor polypeptides into the Sec61 complex; (ii) it binds to the incoming precursor polypeptide, acting as a molecular ratchet; and (iii) it is involved in closing the Sec61 channel, thus preventing the uncontrolled efflux of calcium ions from the ER. The two alternatively acting nucleotide exchange factors of BiP, Sil1 and Grp170, guarantee efficient loading of BiP with ATP. In *S. cerevisiae*, Sec63p is part of a heterotetrameric complex that also includes Sec62p (Table 1). In mammals, Sec63 also associates with Sec62, but an analogous tetrameric complex has not been observed. However, in humans, there is a second J-domain-containing protein present in the ER membrane, termed ERj1, that contacts translating ribosomes on the ER surface (Fig. 1).[16] ERj1 may provide a functional alternative to Sec63. The fact that human ERj1 can complement a yeast Sec63p mutant strain also supports this view.[17]

Recent work on protein translocation across and protein insertion into the ER membrane has demonstrated that different precursor polypeptides have different requirements for components of the protein translocation machinery. These differences became obvious from the identification of selective inhibitor sensitivities as well as from the differential involvement of additional auxiliary proteins, such as TRAM protein, RAMP4 and the TRAP complex.[18-21]

Protein translocation into the ER is essential for cell proliferation. Therefore, it does not come as a surprise that several of the genes coding for proteins involved in translocation are essential for *S. cerevisiae* (Table 1); the same is expected to be the case in humans, too. Driving home this point is the recent observation that a bacterial cytotoxin inactivates BiP and therefore causes life-threatening complications (Fig. 2).[22,23] However, some components of the ER-resident protein translocase have overlapping or redundant functions, making them nonessential for *S. cerevisiae* (Table 1). Therefore, human diseases are expected to exist that originate in genes coding for redundant components of the human ER-resident protein translocase. So far, two examples of such inherited diseases have been discovered (Table 1).[24-28]

A Bacterial Cytotoxin That Inactivates BiP Causes Haemolytic Uraemic Syndrome (HUS)

Shiga toxigenic *Escherichia coli* (STEC) strains cause massive global morbidity and mortality, particularly, in developing countries.[22,23] Typically, these pathogens produce Shiga toxin and AB_5 toxin (also termed subtilase AB) and are responsible for gastrointestinal diseases, such as the life-threatening haemolytic uraemic syndrome (HUS). SubAB is a particularly nasty AB_5 cytotoxin and was responsible for an outbreak of HUS in Australia in 1998. The B-subunit pentamer binds to toxin receptors on the cell surface, while the A-subunit has subtilase-like serine protease activity. During a STEC infection, the bacterial cytotoxin enters human cells by endocytosis, followed by retrograde transport to the ER. In the ER, BiP is the major target of the catalytic A-subunit, which

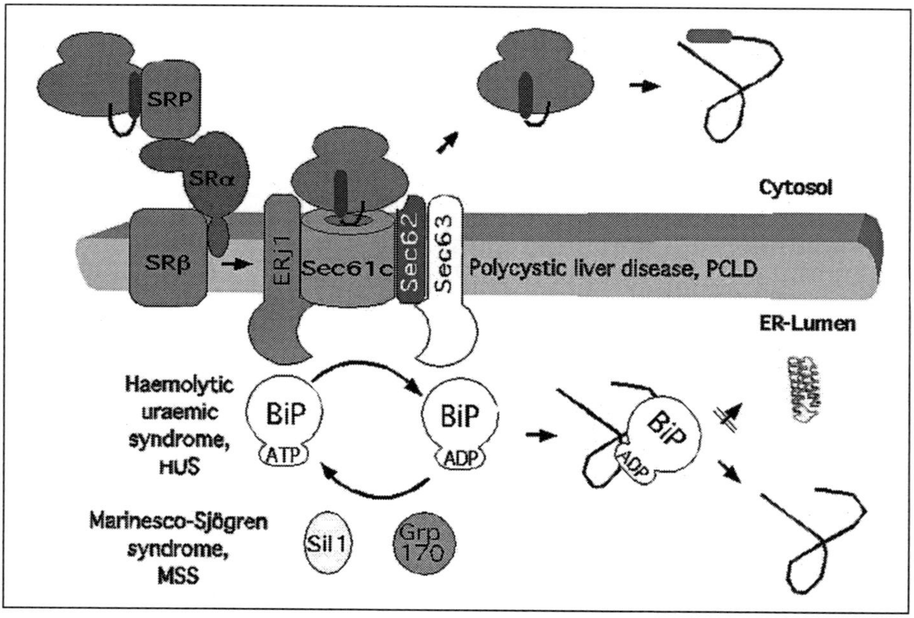

Figure 2. Potential pathologies of protein transport into the endoplasmic reticulum. See text for details. The figure was adapted from reference 33.

inactivates BiP by limited proteolysis. Eventually, all BiP functions, such as protein transport into the ER and protein folding in the ER, are completely lost and the affected cells die (Fig. 2).

Mutations in *SEC63* Cause Autosomal Dominant Polycystic Liver Disease (ADPLD)

Autosomal dominant polycystic liver disease (ADPLD) is a rare human inherited disease characterized by the progressive development of multiple biliary epithelial liver cysts.[24-26] It usually remains asymptomatic until it manifests between the ages of 40 and 60 years. Typically, liver function is preserved, even in advanced cases. However, the mass effect of the enlarged liver can result in symptoms such as abdominal pain and shortness of breath and may eventually require surgical intervention.

ADPLD involves several genes. So far, two have been identified: *PRKCSH (protein kinase C substrate 80K-H)* on chromosome 19q13.2, encoding the ß-subunit of glucosidase II (also termed hepatocystin) and *SEC63* on chromosome 6q21. In *SEC63*, two frame-shift mutations with premature chain termination and three nonsense mutations have been described (Fig. 3A). In addition, a single amino acid substitution as well as a single amino acid deletion have been described and are of particular interest from the functional point of view. A loss of Sec63 function has been postulated in all of these cases (Fig. 2).

Although no mechanism has been firmly established for ADPLD, the disease can be explained by a two-step process (termed two-hit mechanism): patients with one inherited mutant allele and one wild-type allele may lose the wild-type allele in some liver cells through somatic mutation. A plausible scenario is that Sec63 is essential for importing and/or folding of a subset of nonessential secretory or plasma membrane proteins, most likely glycoproteins, that are involved in control of biliary cell growth and proliferation. Thus, without functional Sec63, these proteins do not reach their correct location at the cell surface. The result is uncontrolled proliferation of the respective liver cells, i.e., cyst formation.

Mutations in *SIL1* Cause Marinesco-Sjögren Syndrome (MSS)

Marinesco-Sjögren syndrome (MSS) is a rare autosomal recessively inherited neurodegenerative disease.[27-32] The hallmarks of MSS are cerebellar ataxia, cataracts, developmental and mental retardation and progressive myopathy.

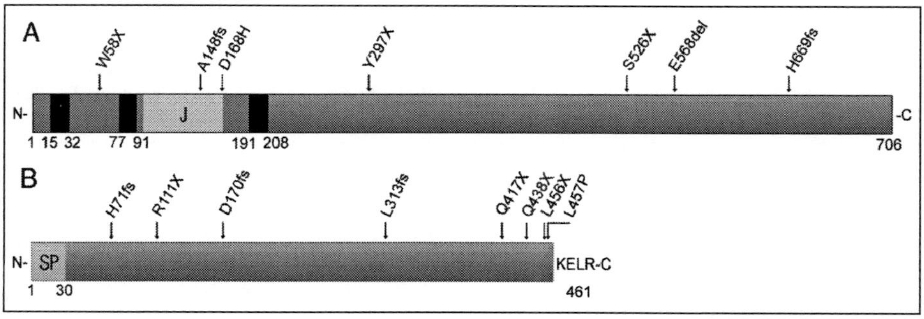

Figure 3. Mutations in *SEC63* and *Sil1* that cause human diseases. The amino acid sequences of the two proteins Sec63 (ERj2) and Sil1 (BAP) are schematically depicted. Special features of the proteins are highlighted, such as the J-domain (J) plus the transmembrane domains (black areas) in the case of Sec63 (A) and the signal peptide (SP) plus the putative ER-retention peptide (KELR) in the case of Sil1 (B). Reported mutations and their effects at the protein level are indicated.[24,26,30-32] Amino acid residues are indicated in the single letter code; C, carboxyl terminus; del, deletion; fs, frame shift; N, amino terminus; X, stop codon. We note that in addition to the indicated mutations, mutations disrupting splice donor-acceptor sites as well as heterozygous mutations (in MSS) were described.[24,26,30-32]

The cause of the disease in a major proportion of MSS patients has been characterized as a mutation in the *SIL1* gene on chromosome 5q31 that results in mutated or truncated Sil1 (Fig. 3B). Thus far, three frame-shift mutations with premature chain termination and four nonsense mutations have been described (Fig. 3B). A loss of Sil1 function was postulated in these cases (Fig. 2). In addition, a missense mutation was described that is of particular interest from the cell biology point of view. This latter mutation (L457P) as well as one of the nonsense mutations (L456X) can be expected to affect ER-retention of Sil1, since the putative retention motif (KELR) is missing from or may not be properly folded in the mutant protein (Fig. 3B).

Sil1 is a nucleotide exchange factor for BiP and its role is to provide ATP-loaded BiP. A loss of Sil1 function, therefore, results in a reduction of functional BiP. There are several possible consequences: (i) some precursor proteins may not be transported into the ER, causing precursor polypeptides to accumulate in the cytosol; (ii) some proteins that are successfully transported into the ER may not be folded correctly, leading to accumulation of misfolded polypeptides in the ER; or (iii) as a consequence of (i) or (ii), some essential secretory or plasma membrane proteins may not reach their functional location, thereby, affecting cell and organ development. Alternatively, overwhelming protein accumulation in the cytosol or the ER lumen could result directly in cell death.

Open Questions

For both of these inherited diseases, future genetic work will undoubtedly identify additional genes that can cause the respective disease and that thus will give further hints about the aetiology of each disease. Furthermore, future studies will have to identify the proteins that are missing from the affected cells and that therefore may cause the disease. In the case of MSS, it will be interesting to see whether it is actually missing proteins or accumulation of aggregated proteins that causes the disease. Further questions concern the reasons why only certain organs are affected in both diseases.

Acknowledgements

The authors' work was funded by the Deutsche Forschungsgemeinschaft.

References

1. Palade G. Intracellular aspects of protein synthesis. Science 1975; 189:347-358.
2. Blobel G, Dobberstein B. Transfer of proteins across membranes. I. Presence of proteolytically processed and unprocessed nascent immunoglobulin light chains on membrane-bound ribosomes of murine myeloma. J Cell Biol 1975; 67:835-851.
3. Zimmermann R, Sagstetter M, Lewis MJ et al. Seventy-kilodalton heat shock proteins and an additional component from reticulocyte lysate stimulate import of M 13 procoat protein into microsomes. EMBO J 1988; 7:2875-2880.
4. Wiedmann B, Sakai H, Davis TA et al. A protein complex required for signal-sequence-specific sorting and translocation. Nature 1994; 370:434-440.
5. Craven RA, Egerton M, Stirling CJ. A novel Hsp70 of the yeast ER lumen is required for the efficient translocation of a number of protein precursors. EMBO J 1996; 15:2640-2650.
6. Dierks T, Volkmer J, Schlenstedt G et al. A microsomal ATP-binding protein involved in efficient protein transport into the mammalian endoplasmic reticulum. EMBO J 1996; 15:6931-6942.
7. Rothblatt JA, Deshaies RJ, Sanders SL et al. Multiple genes are required for proper insertion of secretory proteins into the endoplasmic reticulum in yeast. J Cell Biol 1989; 109:2641-2652.
8. Meyer H-A, Grau H, Kraft R et al. Mammalian sec61 is associated with sec62 and sec63. J Biol Chem 2000; 275:14550-14557.
9. Tyedmers J, Lerner M, Bies C et al. Homologs of the yeast sec complex subunits sec62p and sec63p are abundant proteins in dog pancreas microsomes. Proc Natl Acad Sci USA 2000; 97:7214-7219.
10. Dudek J, Greiner M, Müller A et al. ERj1p plays a basic role in protein biogenesis at the endoplasmic reticulum. Nat Struct Mol Biol 2005; 12:1008-1014.
11. Tyson JR, Stirling CJ. LHS1 and SIL1 provide a lumenal function that is essential for protein translocation into the endoplasmic reticulum. EMBO J 2000; 19:6440-6452.
12. Steel GJ, Fullerton DM, Tyson JR et al. Coordinated activation of Hsp70 chaperones. Science 2004; 303:98-101.

13. Chung KT, Shen Y, Hendershot H. BAP, a mammalian BiP associated protein, is a nucleotide exchange factor that regulates the ATPase activity of BiP. J Biol Chem 2002; 277:47557-47563.
14. Weitzmann A, Volkmer J, Zimmermann R. The nucleotide exchange factor activity of Grp170 may explain the nonlethal phenotype of loss of sil1 function in man and mouse. FEBS Lett 2006; 580:5237-5240.
15. Römisch K. Endoplasmic reticulum-associated degradation. Annu Rev Cell Dev Biol 2005; 21:435-456.
16. Blau M, Mullapudi S, Becker T et al. ERj1p uses a universal ribosomal adaptor site to coordinate the 80 S ribosome at the membrane. Nat Struct Mol Biol 2005; 12:1015-1016.
17. Kroczynska B, Evangelista CM, Samant SS et al. The SANT2 domain of murine tumor cell DnaJ-like protein 1 human homologue interacts with α_1-antichymotrypsin and kinetically interferes with its serpin inhibitory activity. J Biol Chem 2004; 279:11432-11443.
18. Garrison JL, Kunkel EJ, Hegde RS et al. A substrate-specific inhibitor of protein translocation into the endoplasmic reticulum. Nature 2005; 436:285-289.
19. Besemer J, Harant H, Wang S et al. Selective inhibition of cotranslational translocation of vascular cell adhesion molecule 1. Nature 2005; 436:290-293.
20. Voigt S, Jungnickel B, Hartmann E et al. Signal sequence-dependent function of the TRAM protein during early phases of protein transport across the endoplasmic reticulum membrane. J Cell Biol 1996; 134:25-35.
21. Fons RD, Bogert BA, Hegde RS. Substrate-specific function of the translocon-associated protein complex during translocation across the ER membrane. J Cell Biol 2003; 160:529-539.
22. Paton AW, Beddoe T, Thorpe CM et al. AB5 subtilase cytotoxin inactivates the endoplasmic reticulum chaperone BiP. Nature 2006; 443:548-552.
23. Montecucco C, Molinari M. Death of a chaperone. Nature 2006; 443:511-512.
24. Davila S, Furu L, Gharavi AG et al. Mutations in SEC63 cause autosomal dominant polycystic liver disease. Nat Gen 2004; 36:575-577.
25. Drenth JPH, Martina JA, van de Kerkhof R et al. Polycystic liver disease is a disorder of cotranslational protein processing. TREND Mol Med 2005; 11:37-42.
26. Waanders E, te Morsche RHM, de Man RA et al. Extensive mutational analysis of PRKCSH and SEC63 broadens the spectrum of polycystic liver disease. Hum Mutat 2006; 27:830-839.
27. Anttonen A-K, Mahjneh I, Hämälainen RH et al. The gene disrupted in Marinesco-Sjögren syndrome encodes SIL1, an HSPA5 cochaperone. Nat Gen 2005; 37:1309-1311.
28. Senderek J, Krieger M, Stendel C et al. Mutations in sil1 cause Marinesco-Sjögren syndrome, a cerebellar ataxia with cataract and myopathy. Nat Gen 2005; 37:1312-131314.
29. Zoghbi HY. SILencing misbehaving proteins. Nat Gen 2005; 37:1302-11303.
30. Karim MA, Parsian AJ, Cleves MA et al. A novel mutation in BAP/SIL1 gene causes Marinesco-Sjögren syndrome in an extended pedigree. Clin Genet 2006; 70:420-423.
31. Eriguchi M, Mizuta H, Kurohara K et al. Identification of a new homozygous frameshift insertion mutation in the Sil1 gene in 3 japanese patients with Marinesco-Sjögren syndrome. J Neurol Sci 2008; 16:961-969.
32. Anttonen A-K, Siintola E, Tranebjaerg L et al. Novel SIL1 mutations and exclusion of functional candidate genes in Marinesco-Sjögren syndrome. Eur J Hum Genet 2008; 16:961-969.
33. Zimmermann R, Müller L, Wullich B. Protein transport into the endoplasmic reticulum; mechanisms and pathologies. Trends Mol Med 2006; 12:567-573.

Index

T

U

V

X

Y